Max-Olivier Hongler

Chaotic and Stochastic Behaviour in Automatic Production Lines

Springer-Verlag Berlin Heidelberg GmbH

Author

Max-Olivier Hongler
Département de Microtechnique
Ecole Polytechnique Fédérale de Lausanne
CH-1015 Lausanne, Switzerland

These notes basically constitute the contents of a series of lectures first delivered in the 1992-1993 Winter semester at the Bielefeld-Bochum Stochastik Zentrum and Fakultät für Physik of the University of Bielefeld (Germany).
This work has in part been supported by the Stifterverband für die Deutsche Wissenschaft, Essen.

ISBN 978-3-662-14508-1   ISBN 978-3-540-48448-6 (eBook)
DOI 10.1007/978-3-540-48448-6

CIP data applied for.

© Springer-Verlag Berlin Heidelberg 1994

Originally published by Springer-Verlag Berlin Heidelberg New York in 1994
Softcover reprint of the hardcover 1st edition 1994

This book was processed using the LATEX macropackage with LMAMULT style
SPIN: 10080264   55/3140-543210 - Printed on acid-free paper

Lecture Notes in Physics

New Series m: Monographs

The Editorial Policy for Monographs

The series Lecture Notes in Physics reports new developments in physical research and teaching - quickly, informally, and at a high level. The type of material considered for publication in the New Series m includes monographs presenting original research or new angles in a classical field. The timeliness of a manuscript is more important than its form, which may be preliminary or tentative. Manuscripts should be reasonably self-contained. They will often present not only results of the author(s) but also related work by other people and will provide sufficient motivation, examples, and applications.

The manuscripts or a detailed description thereof should be submitted either to one of the series editors or to the managing editor. The proposal is then carefully refereed. A final decision concerning publication can often only be made on the basis of the complete manuscript, but otherwise the editors will try to make a preliminary decision as definite as they can on the basis of the available information.

Manuscripts should be no less than 100 and preferably no more than 400 pages in length. Final manuscripts should preferably be in English, or possibly in French or German. They should include a table of contents and an informative introduction accessible also to readers not particularly familiar with the topic treated. Authors are free to use the material in other publications. However, if extensive use is made elsewhere, the publisher should be informed. Authors receive jointly 50 complimentary copies of their book. They are entitled to purchase further copies of their book at a reduced rate. As a rule no reprints of individual contributions can be supplied. No royalty is paid on Lecture Notes in Physics volumes. Commitment to publish is made by letter of interest rather than by signing a formal contract. Springer-Verlag secures the copyright for each volume.

The Production Process

The books are hardbound, and quality paper appropriate to the needs of the author(s) is used. Publication time is about ten weeks. More than twenty years of experience guarantee authors the best possible service. To reach the goal of rapid publication at a low price the technique of photographic reproduction from a camera-ready manuscript was chosen. This process shifts the main responsibility for the technical quality considerably from the publisher to the author. We therefore urge all authors to observe very carefully our guidelines for the preparation of camera-ready manuscripts, which we will supply on request. This applies especially to the quality of figures and halftones submitted for publication. Figures should be submitted as originals or glossy prints, as very often Xerox copies are not suitable for reproduction. In addition, it might be useful to look at some of the volumes already published or, especially if some atypical text is planned, to write to the Physics Editorial Department of Springer-Verlag direct. This avoids mistakes and time-consuming correspondence during the production period.

As a special service, we offer free of charge LaTeX and TeX macro packages to format the text according to Springer-Verlag's quality requirements. We strongly recommend authors to make use of this offer, as the result will be a book of considerably improved technical quality. The typescript will be reduced in size (75% of the original). Therefore, for example, any writing within figures should not be smaller than 2.5 mm.

Manuscripts not meeting the technical standard of the series will have to be returned for improvement.

For further information please contact Springer-Verlag, Physics Editorial Department II, Tiergartenstrasse 17, D-69121 Heidelberg, FRG.

Contents

Contents

1 Introduction

To provide a unified and fully systematic approach to the study and modeling of the generalized automatic production line (APL) is an almost impossible challenge. Indeed, each production line is most often a custom prototype devoted to a specific product. However, recently, emphasis has been placed on the design and definition of a number of generally applicable guidelines for production systems. The ever-growing complexity and, in consequence, the costs of production lines have stimulated the search for fabrication systems with more *flexibility*. By this concept, we mean the capability of using the same system to produce a family of products rather than just a single one. These flexibility requirements generate several new problems, both at the conception, and at the operating policy levels. In the crowded world of assorted devices, various prototypes, fancy tricks and black magic which has traditionally populated the field of automatic production systems, this additional flexibility requisite imposes the difficult task of identifying those existing devices capable of being adapted, and the building new machines able to work on families of objects. To progress towards this goal, it is necessary to fully understand and control the dynamical behaviours of APLs and ultimately attempt to partly bridge the gaps between the practical know-how and the theoretical research communities. It is obviously very difficult to formulate rigorously the manner in which this reflexion should be conducted. However, it is safe to draw analogies and consider any operating production line as an off-equilibrium thermodynamical system. Fluxes of matter, energy and information are maintained and enable a local diminution of the entropy which results in the occurrence of the required finished products. The thermodynamical equilibrium is reached only when the production is stopped. This very general view reveals that the study of production systems consists in fact of the analysis of the fluxes which drive and maintain the system out-of-equilibrium. In particular, the mechanisms of generation, the amplitudes and the fluctuations of these fluxes completely characterize the dynamics and thereby the performance of a production process, (i.e. the throughput of the APL). Generally, a production line is composed of work-heads, (i.e. machines, manipulators, robots, etc...), which are linked together by transport systems (i.e. conveyors belts, pallets, etc.). The main flux of parts and matter visit these work-heads sequentially, the order being defined by the assembly plan of the final product. Besides this main stream, there are always secondary fluxes of parts which feed the work-heads. Feeding a work-head can itself be a delicate operation. In the most general case, which is unfortunately that most often encountered, the feeding mechanism has to extract from a stock of randomly oriented parts and convey these to a precise location with the required orientation. Moreover, these operations have to be achieved with high and relatively stable fluxes (typically a few parts per second). The generating mechanisms of these feeding fluxes have been classified into three levels in [FI], see Figure 1.1.

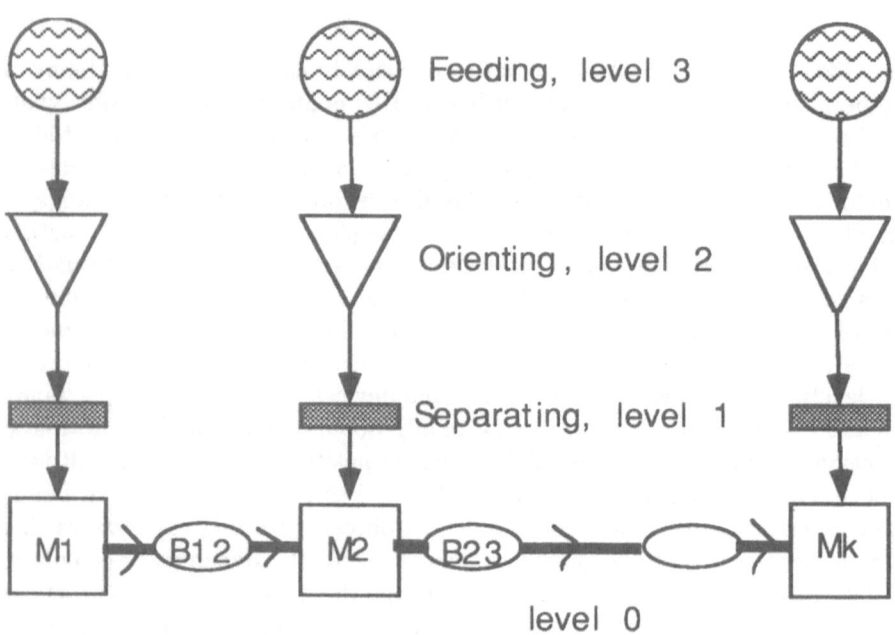

Fig. 1.1. General architecture of an automatic production line. M_k are the work-heads, $B_{k,k+1}$, the buffers. The arrows and the lines stand for the transportation mechanisms.

Each level is characterized by a degree of information entropy that can be defined with the position and orientation of the parts to be conveyed. Level 3, describes the mechanisms which extract parts from a stock. At this level, parts are in a disordered state or pile (random spatial positions and/or orientations), and the feeder is conceived to generate a linear flux of randomly oriented parts. At level 2, one particular orientation is filtered out by the introduction of part orienting and selecting mechanisms. At level 1, the separators assign a referential frame to each correctly oriented part. This referential is used by the downstream work-head to grasp the part and to attach it to the product being assembled flowing at level 0 (i.e. the main flux). Hence, the work-heads operate at the level 0 in this scheme. For each of these levels, a vast array of tools have been developed by the production engineers ([FI], [BO], [CH] and the numerous references therein). At the heart of these lecture notes is the study of the dynamics of a selection of the tools most common to production processes. Even when operating on a purely macroscopic scale (i.e. molecular agitation is irrelevant), random phenomena play a major role in the evolution of the devices discussed. In fact, fluctuations greatly influence the way the production lines are conceived. This will be even more so in the future, when automatic production tools will process micro-systems at scales comparable to those at which the Brownian motion was first discovered. Coming back to the flexibility requisite, it is important to point out the relevant control parameters on which the fluxes of parts depend. The modification of the physical dimensions of some components or modifications in the precision requirements are the typical situations under which the production system should definitely offer good flexibility. These modifications induce variations of the control parameters and possibly of the underlying dynamical system itself, which should leave the behaviour of the line almost unchanged. Therefore the flexibility requirement introduces naturally the notion of generic stability in the operating modes of APLs. Non-linearities, fluctuations and generic stability belong to the basic concepts behind the recent developments of the dynamical systems theory. We are fully convinced that these theoretical concepts should help to systematize the study of production systems. The present text is intended as a start in the exploration of this viewpoint.

2 Vibratory Feeding

Feeding parts, initially oriented randomly, to a manipulator is one of the most commonly found operations in automatic production lines. This is indeed not an easy task as the work-heads are always required to be fed with parts in a specific orientation. To appreciate this, imagine the situation where screws are to be conveyed to an automatic screw-driver. The screws are furnished in open boxes and therefore have no precise orientation. The automatic screw-driver can accept screws only with their heads up. Moreover the feeding of the screw driver has to be performed at high rates, typically a few parts per second. A feeding mechanism is then constructed to decrease the configuration entropy which characterizes the random orientations the parts can assume. This is a basic example of an off-equilibrium thermodynamical system in which energy is supplied in order to reduce the entropy. The feeding devices are usually composed of two distinct parts, the conveying device and the orienting mechanism. One common solution to the conveying problem is provided by vibratory feeders. These machines belong to the level 3 of the general architecture of systems described in the last section (Figure 1.1). Basically, a vibratory feeder consists of an oscillating track on which the parts are conveyed. When the track is set vibrating, the mobile lying on it, is itself set into motion. For properly tuned vibrations, a stationary flux of parts traveling along the track is established. At the end of the track, the system is equipped with orienting and selecting devices. These devices are placed at level 2 of the general configuration sketched in Figure 1.1. The orienting devices ensure that the parts leave the feeder in the required orientation before they enter into the separator situated at level 1. In this chapter, we focus our attention on the conveying mechanism; the orienting and selecting devices will be studied in chapter 3.

The dynamics of feeders recently has been the subject of intensive investigations in mechanical engineering literature [RE], [IN], [JI], [OK1], [TA], [VE], [VA], [HO1], [CA]. One of the pioneering works devoted to this topic is that proposed by A. H. Redford and G. Boothroyd [RE]. While vibratory feeders are in fact tailored for each specific application, this should not prevent us from investigating their common underlying dynamics. To guide our discussion and before we enter into the analysis, let us briefly enumerate a few simple questions relevant to the control of the operation of a vibratory feeder.

a) What are the most important parameters governing the dynamics?

b) How does one estimate the feeding rate?

c) When is the use of a vibratory feeder inadvisable?

d) What is the role played by the form of the excitation? Are these excitations optimal to ensure a high and constant rate?

The answers to this set of questions are usually given by the practical know-how of the constructor. In the following, we will try to obtain some insights into the replies to these questions which result from a mathematical modeling.

2.1 The Basic Dynamical Model

Vibratory feeders can either be bowl shaped or linear. A schematic view of the bowl version of these apparatus is given in Figure 2.1.

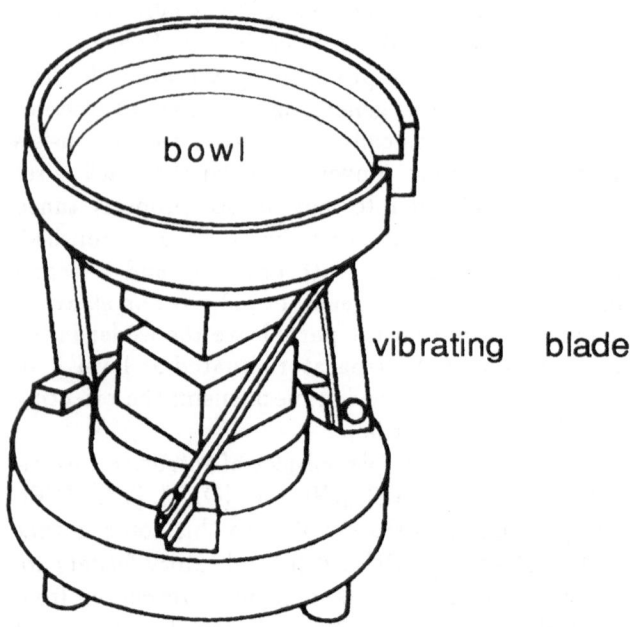

Fig. 2.1. Typical realization of a vibrating bowl feeder. The bowl is mounted on three vibrating blades. The excitations are generated by an electromagnet placed under the bowl.

For the considerations to be made here, we shall take as an approximation the motion of a single part on a purely linear track. The effects due to the angular velocity on helicoidal tracks are very small (i.e. centripetal and Coriolis accelerations are neglected). The part will be assumed to be point-like and not rolling on the track. We shall also neglect the drag due to the air.

The basic forces governing the dynamics are represented on Figure 2.2.

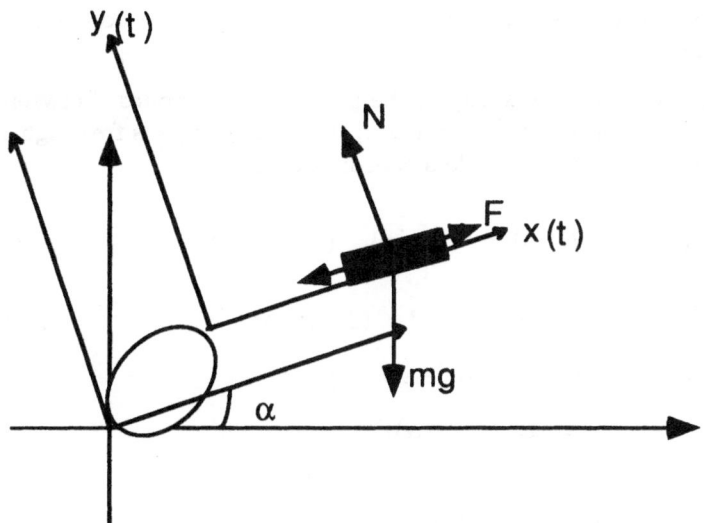

Fig. 2.2. Schematic representation of the vibrating track with a part on it.

The equation of the motion for a part is written in the co-ordinates relative to the track. Accordingly, we have:

$$m\frac{\partial^2}{\partial t^2}\left(x(t)+h(t)\right) = -mg\sin(\alpha) + F_r(t) \qquad (2.1)$$

$$m\frac{\partial^2}{\partial t^2}\left(y(t)+f(t)\right) = -mg\cos(\alpha) + N(t), \qquad (2.2)$$

where $F_r(t)$ and $N(t)$ are respectively the friction and constraint forces, α is the slope angle of the track, g is the gravitation acceleration $h(t) = h(t+P)$ and $f(t) = f(t+P)$, are the positions of vibrating track of the feeder, (P is the period). In actual realizations, [RE], we often have $h(t+\phi) = f(t)$ both f and h being harmonic functions and ϕ representing a phase factor itself commonly taken to be zero.

Depending on the strength of the excitations, the part can either rest on the track, slide, hop above the track or exhibit a mixture of these motions. The friction coefficient plays a central role. It controls the sliding motion in the x direction and the restitution of the tangential and perpendicular velocities after impacts. Of course, the friction coefficient depends on the material used and on external parameters such as the degree of humidity, electrostatic forces, etc... Such external factors can considerably affect the reliability of the system, for instance a feeding rate may change according to the seasons, (i.e. the friction coefficient differs depending on whether it is wet in winter or dry in summer !!).

In this study, we shall assume that the friction coefficient on which the friction force F_r depends, is always high and feeders are likely to operate properly when this is guaranteed. For high F_r, the sliding of parts will remain small. Therefore, the dominant contribution to the transport rate will be due to the hopping of the parts on the tracks.

At present we shall focus on the study of the pure hopping modes. Between the impacts, which occur at times $\{t_k\}, k = 1, 2...$, the motion type is free flight. Hence, Eqs. (2.1) and (2.2) in dimensionless coordinates give:

$$\frac{\partial^2}{\partial \tau^2} \left(q(\tau) + zH(\tau) \right) = -\tan(\alpha), \quad (\| \text{ motion}) \tag{2.3}$$

$$\frac{\partial^2}{\partial \tau^2} \left(u(\tau) + wF(\tau) \right) = -1, \quad (\perp \text{ motion}) \tag{2.4}$$

with the following definitions of dimensionless variables:

$$u = y\omega^2 \left(g\cos(\alpha) \right)^{-1}$$

$$q = x\omega^2 \left(g\cos(\alpha) \right)^{-1}$$

$$w = a\omega^2 \left(g\cos(\alpha) \right)^{-1}$$

$$z = b\omega^2 \left(g\cos(\alpha) \right)^{-1}, \tag{2.5}$$

where a and b are the amplitudes of the first harmonic modes of the Fourier expansion of $f(\tau)$ and $h(\tau)$. The rescaled time τ is denoted by $\tau = \omega t = \frac{2\pi}{P}t = 2\pi\nu t$.

Immediately before and after the impacts (i.e. when $u = 0$), the relative velocities of the parts are related by the following restitution equations:

$$\frac{\partial}{\partial \tau} u(\tau = \tau_k + 0) = -R_\perp \frac{\partial}{\partial \tau} u(\tau = \tau_k - 0) \tag{2.6}$$

and

$$\frac{\partial}{\partial \tau} q(\tau = \tau_k + 0) = R_\| \frac{\partial}{\partial \tau} q(\tau = \tau_k - 0). \tag{2.7}$$

where $R_\perp$ and $R_\|$ are the restitution coefficients. While the content of Eq. (2.6) is obvious, the behaviour described by Eq. (2.7) is only an approximation. Indeed, $R_\|$ should, strictly speaking, depend on the relative change of the normal velocity during the impact. This dependence occurs via the instantaneous effective friction coefficient, [VA]. The approximation implicit in Eq. (2.7) expresses the fact that the friction decelerates the tangential motion at each impact with an effective rate $R_\parallel$. For very high friction, we can approximately take $R_\parallel \approx 0$. In fact, the consistent use of Eq. (2.7) is valid only for large dissipation, i.e. when $R_\parallel \approx 0$.

Integrating Eq. (2.4) between two successive impacts (i.e. $\tau_k \leq \tau \leq \tau_{k+1}$, $k = 1, 2, ...$) and writing v_k for the perpendicular component of the velocity just after the impact at time $\tau = \tau_k$, we have:

$$\frac{\partial}{\partial \tau}(u(\tau) + wF(\tau)) - v_k - w\frac{\partial}{\partial \tau}F(\tau = \tau_k) = -(\tau - \tau_k), \qquad (2.8)$$

and as $u(\tau_k) = 0$, $\forall k$, a further integration yields:

$$u(\tau) + w(F(\tau) - F(\tau_k)) - v_k(\tau - \tau_k) - w\frac{\partial}{\partial \tau}F(\tau = \tau_k)(\tau - \tau_k) =$$

$$= -\frac{1}{2}(\tau - \tau_k)^2. \qquad (2.9)$$

Similarly, Eq. (2.3) with p_k standing for the parallel component of the velocity just after the impact, yields:

$$\frac{\partial}{\partial \tau}(q(\tau) + zH(\tau)) - p_k - z\frac{\partial}{\partial \tau}H(\tau = \tau_k) = -\tan(\alpha)(\tau - \tau_k). \qquad (2.10)$$

Finally at time $\tau = \tau_{k+1} + 0$, i.e. just after the impact at time $\tau = \tau_{k+1}$, the restitution Eq. (2.7) implies:

$$v_{k+1} = -R_\perp\{v_k - (\tau_{k+1} - \tau_k) - w[\frac{\partial}{\partial \tau}(F(\tau = \tau_{k+1})) - \frac{\partial}{\partial \tau}(F(\tau = \tau_k))]\}, \qquad (2.11)$$

and

$$w(F(\tau_{k+1}) - F(\tau_k)) - w\frac{\partial}{\partial \tau}F(\tau = \tau_k)(\tau_{k+1} - \tau_k) -$$

$$- v_k(\tau_{k+1} - \tau_k) + \frac{1}{2}(\tau_{k+1} - \tau_k)^2 = 0. \qquad (2.12)$$

A similar calculation for the parallel component of the motion, yields:

$$p_{k+1} = R_\parallel\{p_k - \tan(\alpha)(\tau_{k+1} - \tau_k) - z[\frac{\partial}{\partial \tau}(H(\tau = \tau_{k+1})) - \frac{\partial}{\partial \tau}H(\tau = \tau_k)]\}. \qquad (2.13)$$

For this pure hopping regime, the transport property of the feeder will be given simply by:

$$T_{hop} = \frac{1}{\tau_{k+1} - \tau_k}\int_{\tau_k}^{\tau_{k+1}}\frac{\partial}{\partial \tau}q(\tau)d\tau =$$

$$-\frac{\tan(\alpha)}{2}(\tau_{k+1}-\tau_k)-z\frac{(H(\tau_{k+1})-H(\tau_k))}{\tau_{k+1}-\tau_k}+p_k+z\frac{\partial}{\partial\tau}H(\tau=\tau_k). \quad (2.14)$$

2.2 Solution of the Impact Equations

Let us now concentrate on the perpendicular component of the motion and solve Eqs. (2.11) and (2.12) in the following situations:

<u>Periodic motion.</u>

Here, we introduce the the trial solution:

$$\tau_k=\tau_0+2\pi jk, \quad j=1,2,\dots. \text{ and } \quad k=1,2,\dots. \quad (2.15)$$

Eqs. (2.11) and (2.12) immediately yield:

$$v_k=v_{k+1}=v=2\pi j\frac{R_\perp}{1+R_\perp} \quad \text{(which is independent of } F(\tau)) \quad (2.16)$$

and

$$w\frac{\partial}{\partial\tau}F(\tau=\tau_0)=j\pi\frac{1-R_\perp}{1+R_\perp}. \quad (2.17)$$

The parallel component of the motion given by Eq. (2.13), has the form:

$$p_{k+1}=p_k=p=-\frac{2\pi j\tan(\alpha)}{1-R_\parallel}R_\parallel \quad (2.18)$$

and therefore, using Eq. (2.14), the transport rate is:

$$T_{hop}=z\frac{\partial}{\partial\tau}H(\tau=\tau_0)-j\pi\tan(\alpha)\frac{1+R_\parallel}{1-R_\parallel}. \quad (2.19)$$

Eq. (2.19) immediately enables the calculation of the critical slope α_c above which no conveying by pure hopping is possible. Indeed, we have:

$$\alpha_c=\arctan\{\frac{z\frac{\partial}{\partial\tau}H(\tau=\tau_0)(1-R_\parallel)}{j\pi(1+R_\parallel)}\} \quad (2.20)$$

The periodic solution described by Eqs. (2.15) and (2.16) is of course not stable for the full range of admissible parameters $(w, R_\perp)$. To obtain the stability domain, we linearize the dynamics in the vicinity of the periodic solution to obtain the characteristic polynomial which has the form, [GO]:

$$\lambda^2-\lambda\left(1+R_\perp^2+w(1+R_\perp^2)\frac{\partial^2}{\partial\tau^2}F(\tau=\tau_0)\right)+R_\perp^2=0. \quad (2.21)$$

The stability of the periodic solution is realized for the combination of $(w, R_\perp)$ values which leads the roots of Eq. (2.21) to be confined to the interior of the unit circle. This is achieved when:

$$1 - R_\perp^2 > 0,$$

$$w \frac{\partial^2}{\partial \tau^2} F(\tau = \tau_0) < 0,$$

and

$$2(1 + R_\perp^2) + w(1 + R_\perp)^2 \frac{\partial^2}{\partial \tau^2} F(\tau = \tau_0) > 0. \tag{2.22}$$

When w is increased beyond the stability region defined by Eq. (2.22), a cascade of period doubling bifurcations occurs, [LI], [GU], [HL], [TU1], [TU2], [BA1], [WI]. A typical scenario is sketched in Figure 2.3.

The notation $(j\,;\,k)$, where the j and k are integers describes a periodic motion characterized by one impact every j cycles of the track and the impacting pattern is repeated every jk cycles, [BA1]. A few examples of possible motions and their associated $(j\,;\,k)$ characteristics are presented in Figures 2.3 and 2.4.

For fixed j, we denote by $w_k(j)$ the values of w where the bifurcations of hopping regimes occur. One of the most profound results in the wealth of recent discoveries on the behaviour of non-linear dynamical sytems has been the observation that the succession of bifurcation points $w_k(j)$ follow a geometric series with a universal ratio, [GU], namely:

$$\lim_{k \to \infty} \frac{w_k(j) - w_{k-1}(j)}{w_{k+1}(j) - w_k(j)} = 4.66... \tag{2.23}$$

where 4.66... is the celebrated Feigenbaum universal constant, [LI], [GU]. Beyond $w_\infty(j)$, chaotic hopping regimes will take place. It is remarkable that the Feigenbaum constant does not depend on the detailed structure of the non-linearity which governs the dynamics. In our case this enables us to conclude that provided the hopping regimes are established a cascade of period doubling bifurcations, characterized by Eq. (2.23), will be observed for periodic, non-harmonic excitations of the vibrating track. The complete qualitative picture of a cascade of bifurcations can be schematized as shown in Figure 2.5.

The visualization of these motions is made particularly clear when using the software program Bouncing Ball proposed by [TU3]. The Figures 2.3 and 2.4 were drawn using [TU3]. Moreover, the experimental set-up described in [TU2], [ML] enables us to observe a few bifurcation points of the theoretical cascade given by Eq. (2.23).

So far, our discussion has been performed for the case of a general excitation $F(\tau)$ of the vibrating track. Let us now work out some explicit formulae for two particular excitation forces.

Case a) <u>Sinusoidal excitations</u>.

$$F(\tau) = \sin(\tau). \tag{2.24}$$

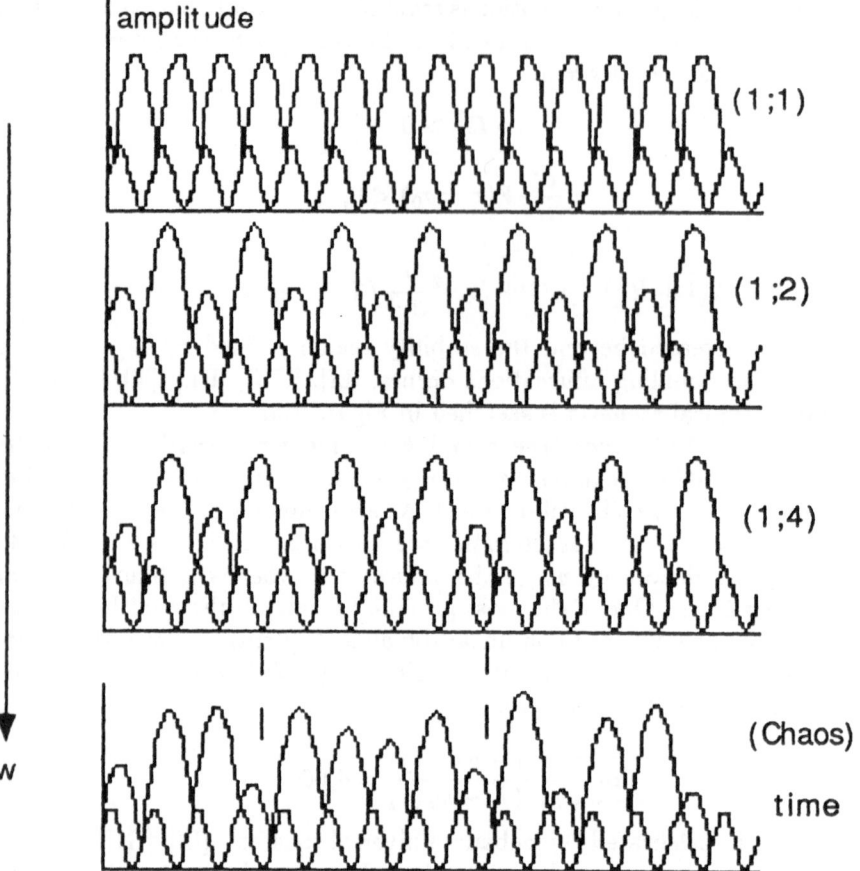

Fig. 2.3. Cascade of period doubling bifurcations followed by the perpendicular component of the motion of the part. The lower oscillations represent the motion of the track, here a sine wave, and the upper trajectories, which are pieces of parabola, are the free flights due to the hopping.

In this case Eq. (2.17) reads:

$$\tau_0 = A\cos\{\frac{j\pi(1 - R_\perp)}{w(1 + R_\perp)}\}. \qquad (2.25)$$

and the stability relations Eq. (2.22) can be written in the form [IN], [BA1], [GU]:

$$j\pi\frac{1 - R_\perp}{1 + R_\perp} \leq w \leq \sqrt{j^2\pi^2\left(\frac{1 - R_\perp}{1 + R_\perp}\right)^2 + 4\frac{(1 + R_\perp^2)^2}{(1 + R_\perp)^4}}. \qquad (2.26)$$

Case b) <u>Piecewise harmonic forces</u> [GO], [CA].

Amplitude

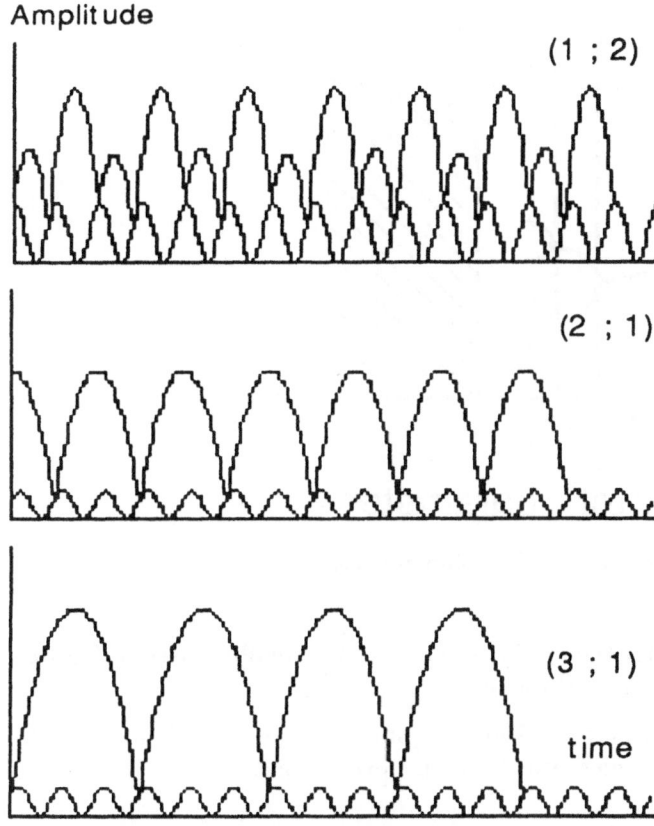

Fig. 2.4. Hopping modes with $(1;2)$, $(2;1)$ and $(3;1)$ characteristics.

$$F(\tau) = \begin{cases} \frac{\pi}{8}\tau(\tau + \pi) & -\pi < \tau \leq 0 \\ \frac{\pi}{8} - \tau(\tau - \pi) & 0 < \tau \leq \pi \end{cases}. \tag{2.27}$$

The amplitude of the first Fourier mode of the excitation given by Eq. (2.27) coincides with the excitation $F\tau = \sin(\tau)$. In this case, we obtain:

$$\tau_0 = \frac{\pi}{2} - \frac{4j(1 - R_\perp)}{w(1 + R_\perp)} = \frac{\pi}{2} - \frac{j\pi(1 - R_\perp)}{x(1 + R_\perp)}, \tag{2.28}$$

where the following notation is used:

$$x = \frac{w\pi}{4}. \tag{2.29}$$

The stability domain is bounded in this case. It is defined by [GO], [BA1]:

$$2j\frac{1 - R_\perp^2}{(1 + R_\perp)^2} \leq x \leq 2\frac{1 + R_\perp^2}{(1 + R_\perp)^2}. \tag{2.30}$$

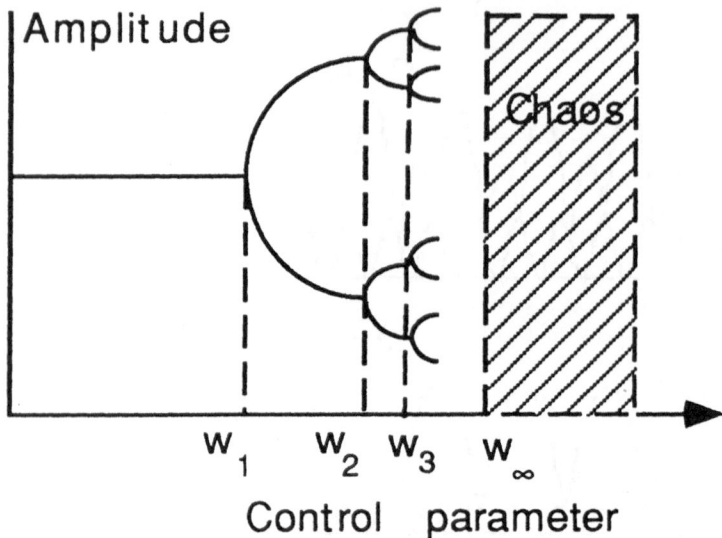

Fig. 2.5. Qualitative picture of a cascade of bifurcations.

Typical stability domains for the $(j\,;\,1)$ hopping modes are illustrated in Figures 2.6 and 2.7 for both types of excitations.

It should be noticed that for $j > 1$ and $R_\perp$ small, stable periodic motions do not necessarily exist for the piecewise harmonic excitations.

2.3 Approximate Mapping for Weakly Dissipative Cases

The above analysis can be made even simpler when $R_\perp$ is not far from unity, (i.e. weakly dissipative dynamics). Although this case is barely encountered in vibratory feeders, we nevertheless briefly present its analysis. When $R_\perp$ is nearly unity, the distance the ball travels between successive impacts becomes large compared to the overall displacement of the table. This property suggests the use of an approximate mapping to describe the dynamics, [BA1]. Introducing the notation:

$$\left(v_{k+1} + w\frac{\partial}{\partial\tau}F(\tau = \tau_{k+1})\right) = \frac{1}{2}\psi_{k+1} \tag{2.31}$$

and assuming:

$$w\left(\frac{F(\tau_{k+1}) - F(\tau_k)}{\tau_{k+1} - \tau_k}\right) \approx 0, \tag{2.32}$$

Eqs. (2.11) and (2.12) become:

$$\tau_{k+1} = \psi_k + \tau_k \tag{2.33}$$

and

$$\psi_{k+1} = R_\perp\psi_k + 2w(1 + R_\perp)\frac{\partial}{\partial\tau}F(\tau = \tau_k + \psi_k). \tag{2.34}$$

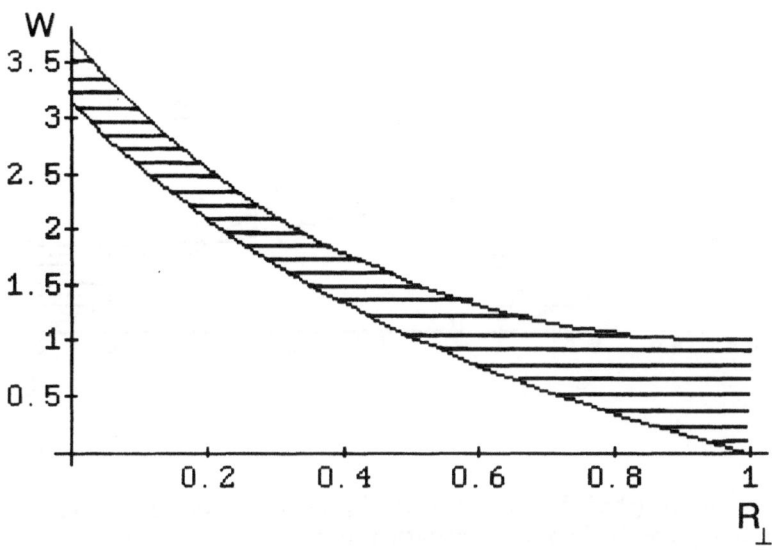

Fig. 2.6. Stability domain (shaded area) given by Eq. (2.26) for the (1; 1) hopping mode resulting from a sinusoidal excitation.

The periodic regime characterized by Eq. (2.15), yields again Eq. (2.17) while we immediately find that $\psi_{k+1} = \psi_k = \psi = 2\pi j$; $j = 1, 2, ...$ The local linearization of the mapping Eqs. (2.33) and (2.34) in the neighbourhood of this solution can be written in the form:

$$\begin{pmatrix} \Delta\tau_{k+1} \\ \Delta\psi_{k+1} \end{pmatrix} = \begin{pmatrix} 1 & 1 \\ \beta\ddot{F} & (R_\perp + \beta\ddot{F}) \end{pmatrix} \begin{pmatrix} \Delta\tau_k \\ \Delta\psi_k \end{pmatrix}, \qquad (2.35)$$

where for brevity, we have adopted the notations:

$$\ddot{F} = \frac{\partial^2}{\partial\tau^2} F(\tau = \tau_k + \psi_k)$$

and

$$\beta = 2w(1 + R_\perp).$$

The characteristic equation associated with Eq. (2.35) is given by:

$$\lambda^2 - \lambda(1 + R_\perp + \beta\ddot{F}) + R_\perp = 0. \qquad (2.36)$$

Hence, we end with:

$$\lambda_\pm = \frac{1}{2}\left(1 + R_\perp + \beta\ddot{F} \pm \sqrt{(1 - R_\perp)^2 + \beta\ddot{F}\left[\beta\ddot{F} + 2(1 + R_\perp)\right]}\right). \qquad (2.37)$$

Therefore, the stability domain is now defined by:

$$w\ddot{F} < 0,$$

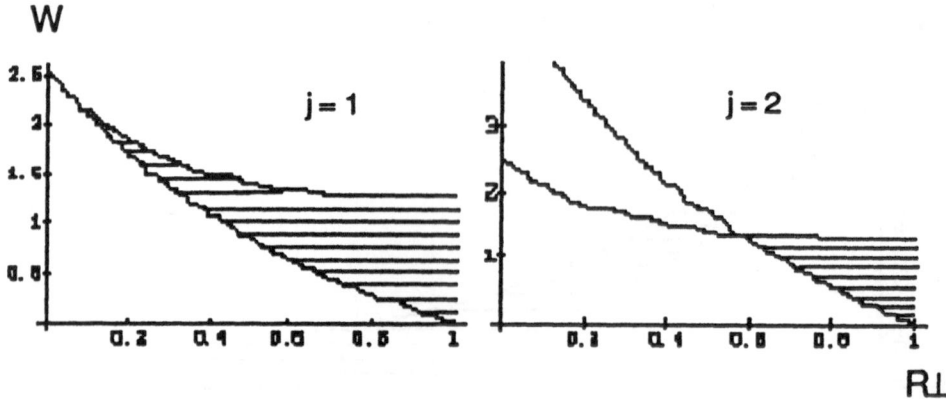

Fig. 2.7. Stability domain (shaded area) given by Eq. (2.30) for the $(j; 1)$ hopping mode resulting from a piecewise harmonic excitation; j=1, 2.

$$w\ddot{F} + 1 > 0$$

or equivalently

$$-1 < w\ddot{F} < 0. \tag{2.38}$$

For the two examples of excitations given by Eqs. (2.24) and (2.27), the stability regions take the forms:

a) <u>Sinusoidal excitation</u>

$$j\pi \frac{1 - R_\perp}{1 + R_\perp} \leq w \leq \sqrt{(j\pi)^2 \frac{(1 - R_\perp)^2}{(1 + R_\perp)^2} + 1}. \tag{2.39}$$

b) <u>Piecewise harmonic excitation</u>

$$0 < w < \frac{4}{\pi}, \text{ independent of } R_\perp. \tag{2.40}$$

When $R_\perp$ is close to unity, a comparison between Eqs. (2.26), (2.30) and Eq. (2.39) and (2.40) respectively, clearly shows that the approximate mapping Eqs. (2.33) and (2.34) yield stability domains closely related to the original mapping, see Figure 2.8.

It is well established that for $R_\perp \approx 1$, the Smale's Horse Shoe (SHS) can be extracted from the mapping Eqs. (2.33) and (2.34), [GU]. The cascade of bifurcations mechanism can be understood directly from the study of dynamical systems in which a (SHS) is embedded [GU], [HL]. For piecewise harmonic excitations Eq. (2.27), the relevant mapping is a 2-D version of the piecewise linear (i.e. also called the tent) map for which numerous exact results can been derived [LA], [HL].

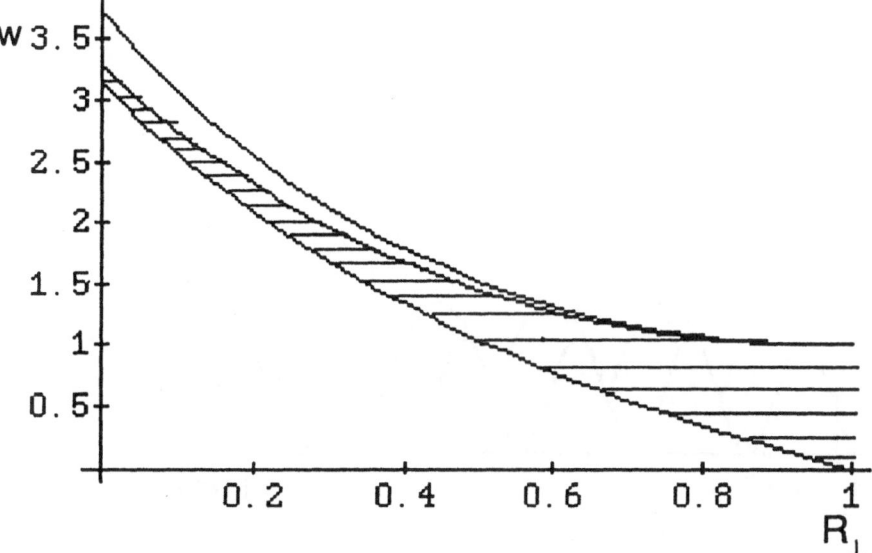

Fig. 2.8. Comparison of the stability domains of the $(1; 1)$ motion obtained with harmonic excitations when the approximate Eq.(2.39) and the exact results Eq. (2.26) are used. The shaded area is the approximate stability domain given by Eq. (2.39). For $R_\perp$ in the vicinity of 1, both domains clearly coincide.

2.4 Plastic Modes

As we already have pointed out, the resting and/or sliding of parts is also observed in the actual motions of the conveyed parts. These modes, called plastic when $R_\perp = 0$ or quasi-plastic (for $R_\perp > 0$) arise when $u = 0$ for a fraction of the period of the excitation [GO], [NA], [KE]. Intuitively, these regimes will be most often observed for strongly dissipative dynamics, (i.e. when $R_\perp \ll 1$). Typical plastic and quasi-plastic modes are respectively sketched in Figure 2.9 and 2.10 when the period of the motion equals the period of the excitation (i.e. $(1; 1)$ modes).

Let us denote by γ the time at which the part leaves the track. Therefore we have:

$$w\ddot{F}(\gamma) + 1 = 0. \tag{2.41}$$

For $\tau > \gamma$, the parts obey the free flight dynamics and we can write:

$$u(\tau) = -w\left(F(\tau) - F(\gamma)\right) - \frac{1}{2}(\tau - \gamma)^2 + w\dot{F}(\gamma)(\tau - \gamma). \tag{2.42}$$

For $R_\perp = 0$, (purely plastic regime), we denote by τ^* the time at which the part lands again on the track. Hence, using Eq. (2. 42), we have:

$$-wF(\tau^*) + wF(\gamma) - \frac{1}{2}(\tau^* - \gamma)^2 + w(\tau^* - \gamma)\dot{F}(\gamma) = u(\tau^*) = 0, \tag{2.43}$$

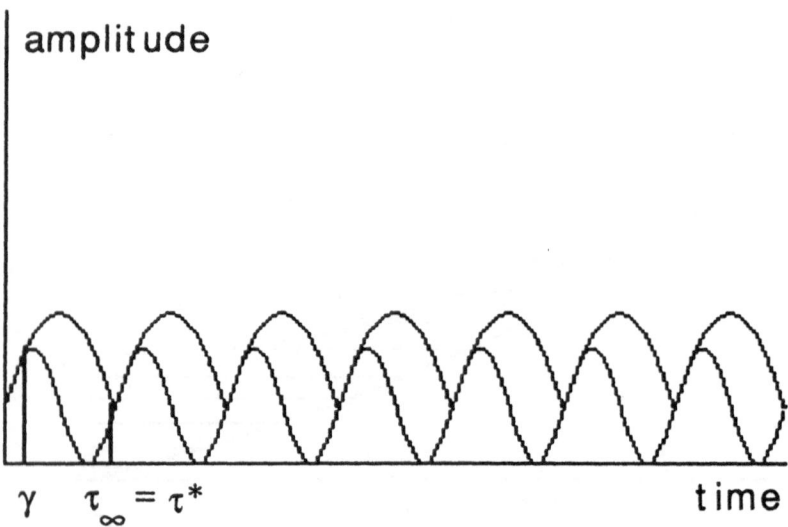

Fig. 2.9. Sketch of the vertical component of the motion for a plastic $(1 ; 1)$ mode. In this plastic case, (i.e. $R_\perp = 0$), the first impact time τ^* and the landing time τ_∞ are identical.

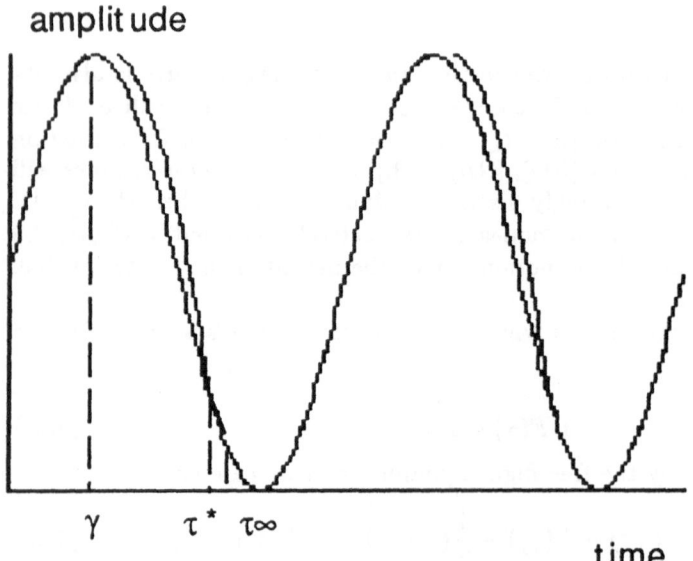

Fig. 2.10. Sketch of the vertical component of the motion for a quasi-plastic $(1 ; 1)$ mode. Here $R_\perp > 0$ and therefore $\tau^* < \tau_\infty$. The successive hoppings which occur after time τ_∞ are too small to be clearly visible on the picture.

with the additional restriction:

$$2\pi j + \gamma' \le \tau^* \le 2\pi j + \gamma, \qquad (2.44)$$

where γ' is the nearest root to γ of the equation $w\ddot{F}(\gamma') + 1 = 0$ for which we have $\gamma' < \gamma$ and $w\ddot{F}(\gamma) > 0$.

When $R_\perp > 0$, quasi-plastic modes are induced. A quasi-plastic mode regime is characterized by the fact that for $\tau > \tau^*$, the part hops an infinite number of times in a finite time interval until it definitely lands on the track at τ_∞, (see Figures 2.9 and 2.10 for the definitions of γ, τ^* and τ_∞.)

To better study these regimes, we focus our attention on the piecewise harmonic excitations Eq. (2.27). In this case, the calculations can be performed explicitly. Here, we immediately obtain $\gamma = 0$ provided:

$$x = \frac{w\pi}{4} > 1. \qquad (2.45)$$

Eq. (2.43) takes the form:

$$-\frac{x}{2}[(\tau^* - \eta\pi)(\tau^* - (\eta - 1)\pi)] - \frac{1}{2}(\tau^*)^2 + \tau^*\frac{\pi x}{2} = 0, \qquad (2.46)$$

with $\eta = 2j$. The positive solution of Eq. (2.46) reads:

$$\tau^* = \pi \left(\frac{\eta x + \sqrt{\eta x(x + 1 - \eta)}}{1 + x} \right). \qquad (2.47)$$

The condition given by Eq. (2.44) implies in this case that $\gamma' = -\pi$ and therefore we obtain:

$$(\eta - 1)\pi \le \pi \left(\frac{\eta x + \sqrt{\eta x(x + 1 - \eta)}}{1 + x} \right) \le \eta\pi,$$

which is equivalent to:

$$\eta - 1 \le x \le \eta. \qquad (2.48)$$

When $R_\perp = 0$ Eq. (2.48) defines the domain of existence for a pure plastic mode.

Let us now consider the quasi-plastic regimes. Introducing the excitation Eq. (2.27) into the impact Eqs. (2.8) and (2.9), we obtain:

$$v_{k+1} = -R_\perp \{ v_k + (\tau_k - \tau_{k+1})(1 + x) \}, \qquad (2.49)$$

and

$$v_k = \frac{1 + x}{2} (\tau_{k+1} - \tau_k). \qquad (2.50)$$

Eqs. (2.49) and (2.50) are solved explicitly by the expressions:

$$v_k = R_\perp^{k+1} v_0, \qquad (2.51)$$

and

$$\tau_k = \tau_0 + \frac{2v_0 R_\perp (1 - R_\perp^k)}{(1+x)(1-R_\perp)},$$ (2.52)

where v_0 and τ_0 are general initial conditions.

At the end of the free flight (i.e. when $\tau = \tau^* - 0$), the velocity of the part is denoted by v^* and is given by:

$$v^* = -\tau^* + \frac{\pi x}{2}.$$ (2.53)

The relative velocity at the moment of the impact is then given by Eq. (2. 25) and reads:

$$v_{\text{rel}} = -\tau^*(1+x) + \eta\pi x.$$ (2.54)

After the impact at time $\tau = \tau^*$, the velocity of the part is:

$$v_{\text{rel}}(\tau = \tau^* + 0) = -R_\perp (v_{\text{rel}}(\tau = \tau^* - 0)) = R_\perp (\tau^*(1+x) - \eta\pi x).$$ (2.55)

Taking $\tau^* = \tau_0$ and $v_{\text{rel}}(\tau = \tau^* + 0) = v_0$ in Eq.(2.52), and letting $k \to \infty$, we obtain:

$$\tau_\infty = \tau^* \left(1 + \frac{2R_\perp^2}{1 - R_\perp}\right) - \frac{2\eta x \pi R_\perp^2}{(1+x)(1-R_\perp)}.$$ (2.56)

Making use for τ^* of the condition given by Eq. (2.44) with $\gamma = 0$ and $\gamma' = -\pi$:

$$\pi(\eta - 1) \le \tau_\infty \le \pi\eta$$ (2.57)

we then obtain:

$$(\eta - 1)\pi \le$$

$$\le \pi\{[1 + \frac{2R_\perp^2}{1 - R_\perp}]\frac{\eta x + \sqrt{\eta x(1 + x - \eta)}}{1 + x} - \frac{2\eta x R_\perp^2}{(1+x)(1-R_\perp)}\} \le$$

$$\le \eta\pi,$$ (2.58)

or written differently:

$$(\eta - 1) \le x \le \frac{1}{2}\left((\eta - 1) + \sqrt{(\eta - 1)^2 + \frac{4\eta(1 - R_\perp)^2}{(1 - R_\perp + 2R_\perp^2)^2}}\right),$$ (2.59)

with:

$$\eta = 2j = 2, 4, 6, \ldots.$$

Observe that for $R_\perp = 0$, Eq. (2.59) consistently reduces to the plastic mode condition given by Eq. (2.48). On the other hand, when $R_\perp = 1$, Eq. (2.59) simply implies that $x = \eta - 1$ which, in view of Eq. (2.48), indicates that the stability domain of the quasi-plastic modes is reduced to a single point, (i.e. quasi-plastic modes do not exist as one could have intuitively expected for the

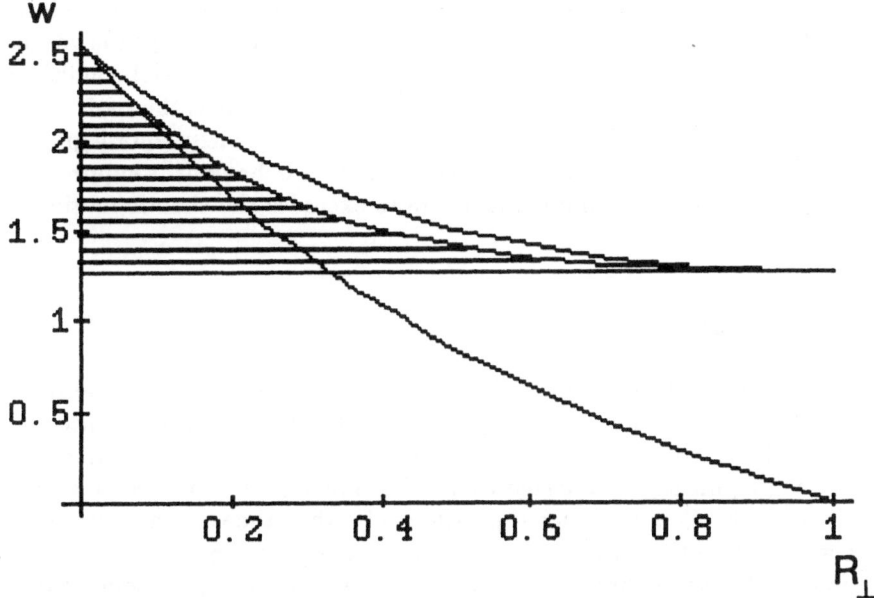

Fig. 2.11. Stability domain, (shaded areaa),of the $(1\,;1)$ quasi-plastic mode for the piecewise harmonic excitations, (Eq. (2.59)). For comparison, the superimposed curves define the stability domain of the $(1\,;1)$ pure hopping mode already sketched in Figure 2.7.

non-dissipative dynamics). A sketch of the domain of stability for the quasi-plastic modes is given in Figure 2.11.

This figure exhibits that for $R_\perp \ll 1$, the width of the stability zone of quasi-plastic motions is much larger than the zone of the purely hopping modes. Hence, for moderate w, stable quasi-plastic modes are largely predominant in the dynamics. When $j > 1$ more complex plastic regimes, compatible with the underlying periodic regimes, are obviously also possible. Again, with the excitation Eq. (2.27) explicit calculations can be performed [GO].

So far, the calculations have been presented only for the piecewise harmonic excitations of Eq. (2.27). However, for small $R_\perp$, the same program can be performed. Indeed, when $R_\perp$ is small, the time interval $[\tau^*, \tau_\infty]$ is short. Accordingly, the excitation force $F(\tau)$ can be approximated locally by a parabola. In other words, we can take $F(\tau) = \hat{\alpha}\tau^2 + \hat{\beta}\tau + \hat{\gamma}$ where, $2\hat{\alpha} = \ddot{F}(\tau^*)$ and $\hat{\beta} = \dot{F}(\tau^*)$. With this approximation and with the replacement of $x = \frac{w\pi}{4}$ by $\hat{x} = w\hat{\alpha}$, Eqs. (2.51), (2.52) and (2.56) remain unchanged. Hence, only the calculation of τ^* remains, which results from the solution of a transcendent equation. Applying this method to the sinusoidal excitations, it can be checked again that the quasi-plastic modes dominate for large dissipation $R_\perp \ll 1$ and moderate w parameters.

It is important to point out here that the presence of quasi-plastic modes excludes the existence of a chaotic regime for the perpendicular motion. Indeed, the systematic relaxation of the $u(\tau)$ component of the motion toward the vibrating track prevents the motion from exhibiting sensitivity to its initial conditions for asymptotically large times. Sensitivity to initial conditions is the basic property shared by all chaotic dynamical systems. Therefore, when (quasi-)plastic modes are present, the possibility of observing chaotic behaviour is suppressed. Clearly, the occurrence of sliding parts therefore excludes the chaotic motion of the hopping modes.

2.5 Transport Rates

In general the transport rate is difficult to calculate in a purely analytical manner. Indeed, both the sliding and hopping modes have to be taken into account as they definitely contribute to the overall transport properties of the feeder. The study of sliding regimes (also referred to as stick and slip motions), is presented in [OK1], [FD], [FE]. Sliding, which is possible only when the vertical component coincides with the motion of the track, can occur before $\tau = \gamma$ or after $\tau = \tau_\infty$, (see section 2.4 for the definition of γ and τ_∞). As we have just pointed out, the fact that the perpendicular component of the motion lies on the track for a portion of the excitation period implies that only periodic motions are sustained. For instance, an interesting dynamical scenario is presented in [CA] where the amplitude of a piecewise harmonic excitation is selected such that the resulting motion of the part is on the onset of the hopping regime. This motion is periodic and enables high transport rates.

It is always possible to calculate separately the transport rate due to the hopping and the sliding portion of the motion of a part. The hopping contribution to the mean transport rate T_{hop} is due to the free flight motion. It can be expressed as:

$$T_{hop} = \frac{1}{\tau^* - \gamma} \int_\gamma^{\tau^*} \frac{\partial}{\partial \tau} q(\tau) d\tau = \frac{q(\tau^*) - q(\gamma)}{\tau^* - \gamma}. \qquad (2.60)$$

For a purely hopping regime in the $(1;1)$ mode, we have $\gamma = \tau_k$ and $\tau_\infty = \tau_{k+1}$. Therefore Eq. (2.60) reduces to Eq. (2.14). In most practical realizations of actual feeders, the normal and tangential excitations are such that:

$$F(\tau) = H(\tau + \phi), \qquad (2.61)$$

where ϕ is a dephasing factor which, when properly chosen, can improve the transport rate, [RE]. Observe for instance that in the $(1;1)$ regime, $T = T_{hop}$ as given by Eq. (2.19) is maximum when $\dot{H}(\tau_0) = \dot{F}(\tau_0 - \phi)$ reaches its largest value. In most situations however, the cost and the reliability of the feeders favours the choice $H(\tau) = F(\tau)$. In [RE], the effect of the introduction of a dephasing factor ϕ is studied. When this is the case and the motion is tuned in the $(j;j)$ mode Eq. (2.17) and (2.19) yield:

$$T = T_{hop} = j\pi\{\frac{z(1 - R_\perp)}{w(1 + R_\perp)} - \tan(\alpha)\frac{1 + R_\parallel}{1 - R_\parallel}\}. \qquad (2.62)$$

In Eq. (2.62), the transport rate T does not depend explicitly on $F(\tau)$. This can be understood from the fact that in a pure hopping mode, the part spends the majority of its time flying above the track. Therefore, it does not feel the detailed and intrinsic structure of the excitations. At first glance, it would appear that this property is essential for guaranteeing the reliability and the stability of the transport rate. Indeed, small alterations, (for instance external noise) to $F(\tau)$ should therefore not seriously affect T. Moreover, the parts suspended mostly in flight are much less influenced by the state or varying conditions at the surface of the track. Hence, humidity, electrostatic properties, etc... should also barely influence the transport rates. Unfortunately, tuning a stable $(j\,;\,j)$ mode is difficult in industrial contexts which requires both robustness under external perturbations and flexibility under modifications of the geometry of parts to be transported. The difficulties originate mainly for the following reasons:

a) When $R_\perp \ll 1$, which is the case in most practical applications, the stability zone of the $(j\,;\,j)$ modes is relatively narrow, (see Figures 2.6 and 2.7). Therefore small fluctuations of the external control parameters may rapidly destabilize the motion and generate strongly fluctuating transport rates.

b) Our present modeling of the motion does not include the geometrical effects inherent to the shape of the parts to be transported, (i.e. the parts were assumed to be point-like). It is obviously difficult to fully take into account these geometric factors in the dynamics. To appreciate the complexity of the problem, note that a part may land on the track in different positions and the restitution coefficient $R_\perp$ cannot be treated as a constant anymore. Instead, $R_\perp$ should merely be taken as a random variable. Let us mention that a study which explicitly introduces a probability distribution for $R_\perp$ has been performed by [WO]. Hence, for a narrow stability domain of the control parameters, $(w, R_\perp)$, the stability of the motion in a $(j\,;\,j)$ mode is delicate to achieve industrially.

It is necessary to find a compromise which includes a large flying time and a wide stability domain in the $(w, R_\perp)$ plane. Furthermore, the dynamics must not be too sensitive to the geometrical forms of the parts. Two possibilities seem to compete; either we select excitations which generate a chaotic motion or we stimulate quasi-plastic modes. It is appealing to try to tune the dynamics to a chaotic regime. Indeed, the resulting stochastic processes might be expected to present stationary properties which are relatively insensitive to alterations of the external control parameters and the geometry of the parts. Moreover, chaotic motions contain numerous high flights of the parts which, in the mean, would lead to high transport rates. Such a point of view has been examined in [VA] in a series of numerical experiments. For stongly dissipative systems, i.e. when $R_\perp \ll 1$, the chaos domain in the $(w, R_\perp)$ plane is localized for relatively strong excitations. When chaotic regimes are selected, the associated large accelerations of the track are likely to cause undesirable secondary effects such as: overcrossing of parts, interference between pieces, undesirable boundary effects (i.e. the heights of the jumps are obviously limitated by the separation gap between two consecutive spires of the helicoidal track), etc... The cumulative effects of these side phenomena will greatly alter the ideal theoretical and numerical predictions. In fact erratic motions have been observed to lead to inefficient transport rates [BO] and should be rejected.

This directs our attention to the quasi-plastic modes which definitely present the operating regimes of most interest.

The sliding contribution to the transport rate is intricate to estimate analytically. A further analysis of the slippage is presented here [OK1], [NA], [KE]. The combination between sliding and hopping can lead to numerous scenarios which are summarized in the diagram Figure 2.12, [BO]. For large friction coefficients however, the transport rate due to the sliding contributions is expected to be small compared to the contribution T_{hop} given by Eq. (2.60).

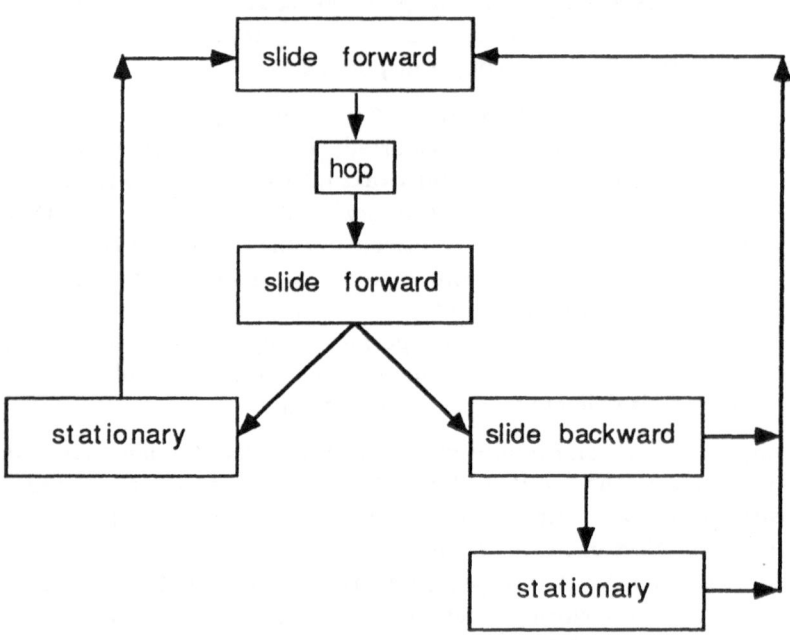

Fig. 2.12. Possible scenario for the complete motion, (i.e. hopping and sliding), of a part on the track.

Our final remark will result from the analysis of the transport rate Eq. (2.62) when it is rewritten with the help of the physical units. Making use of Eq. (2.5), we obtain easily:

$$T_{\text{phys}} = \frac{g\cos(\alpha)}{\omega}T_{hop} = \frac{g\cos(\alpha)}{\omega}T \; ; \quad [\frac{\text{m}}{\text{s}}] \tag{2.63}$$

Hence, for a fixed regime of hopping, reducing the frequency of the vibration increases the transport rate, an observation experimentally reported in [BO]. There is obviously a limit. Indeed, from Eq. (2.5), we see that fixing w (and hence the regime of transport) and decreasing ω implies that we increase the amplitudes of the vibrations and hence the heights of the flights. But these are limited, as we already have observed, by the geometrical constraints inherent to the feeder, in particular the size of the gap between superposed spires of the guiding track.

2.6 Conclusions and Summary

The vibrating feeder is the prototype of a class of mechanical devices capable of behaving erratically even for a purely deterministic equation of the motion. In mechanical engineering vibro-impact devices are commonly encountered. These devices are characterized by the fact that non-linearities are generated by the boundary conditions (impacting dynamical systems). This general context is analyzed in [BA2] and other mechanical engineering texts as for instance described in [HO4], [MV], [HE], [MO] and the numerous references therein.

The mathematical modeling of the vibrating feeder reveals the following features:

a) To avoid a transport rate too sensitive to surface effects (i.e. humidity, electrostatic effects, ...), and too dependent on the fine structure of the excitation forces, it is advantageous to induce a hopping behaviour in the motion of the parts, (i.e. when free-flying, the part is unaware of the existence of the track).

b) Purely hopping modes can be sustained and yield high transport rates. The transport rate is higher for strongly dissipative systems, (i.e. for small restitutions of impulse after the impacts). However, the range of external parameters, (amplitude of excitations and frequency) for which these modes are stable is narrow especially for the highly dissipative processes of the impacting part with the vibrating track. This fine tuning of external parameters leading to stable purely hopping regimes in an industrial context is therefore not feasible.

c) For strongly dissipative situations, a large domain in the admissible external control parameters yields quasi-plastic modes which coexist with stable periodic motions. These motions are characterized by the fact that the parts sit on the track of the feeder for a portion of the excitation cycle. The existence of plastic modes excludes chaotic regimes. This results from the fact that the property of sensitivity to initial conditions, which generates chaotic motions, is destroyed by the relaxation process which causes the part to rest on the track for a portion of the excitation period.

The previous observations suggest that, when hopping is present, there is little hope of drastically improving the feeding rate by introducing fancy excitation forces. Remark however that a simple dephasing factor between the horizontal and vertical components of the excitation can improve the transport rate. One definitely should select the quasi-plastic regimes (i.e. the mixed hopping and sliding regimes). During the hopping portion of the transport, the surface effects and the detailed nature of the excitation force are irrelevant and the landing on the track implies that the dynamics is not sensitive to initial conditions. Moreover, for strongly dissipative systems, $R_\perp \approx 0$ stable quasi-plastic modes exist for a wide range of the excitation parameters. The intentional reduction of the restitution coefficient, by coating the surface of the track [BO], or by the introduction of bristled tracks [OK2], for instance, enlarges the stability domain of the quasi-plastic modes. The reduction of $R_\perp$ also results in an increase in the transport rates. It is therefore important to realize that the quasi-plastic regimes imply a relatively large flexibility in the transport mechanism. Clearly, each time a part rests on the track for a fraction of the excitation period, the initial conditions, which define the characteristics of its free flight, are reset. This systematic re-initialization of initial conditions is performed after each landing on the track. Hence all the fine structure of the trajectories which could result from the internal degrees of freedom (such as the geometrical shape of the parts and other factors omitted by our idealization) are systematically erased. As a result, we end up with a relatively robust and simple motion which will be common to a large variety of components to be transported.

3 Part Orienting Devices

3.1 The General Problem of Selecting Parts in a Correct Orientation

In most feeding situations and especially in the case of feeding small components, the parts are required to arrive to the workheads in identical orientations. In general, this is by no means a simple problem and various solutions can and have been adopted in practice. Here, we present the modeling of a class of mechanisms which rely entirely on the the geometric properties of the parts and that of the gravitational field; these are the Part Orienting and Selecting Systems (POSelSs). These POSelSs are able to continuously select and discriminate, from an incoming flux of randomly oriented parts, those that possess the desired specific and precise orientations. A POSelS is generally located at the end of the track of a vibrating feeder as introduced in chapter 2. Hence, according to the general architecture of systems (see Figure 1.1), a POSelS operates at level 2. The parts leave a POSelS in one required configuration and are then conveyed further downstream to a separating device, (situated itself at level 1 in Figure 1.1), which assigns a referential frame to each of the incoming parts.

A POSelS is generally made of two types of device in series, namely the Part Orienting Device (POD) which precedes the Part Selecting Device (PSelD), which lets through only parts with the required orientation. Clearly, the PSelDs, will reduce the flux of parts (in practice, the parts with the wrong orientations are simply refed to the bottow of the vibrating feeder). A POD, on the other hand, conserves the flux of parts. In order to get a reasonably high flux of correctly oriented parts, it is common to introduce a cascaded series of PODs, which constitute a Part Orienting System (POS), before the flux of parts enters the selection zone, itself composed of one or several PSelDs which act as projectors, (i.e. after leaving the PSelD the probability that the components are in the required orientation is one). A typical POSelS is sketched in Figure 3.1.

It is interesting to note that these devices rely entirely on the presence of a gravitational field. This is even more explicit for the POD known as Gravity Part-Feeder [HN], sketched in Fig. 3.2.

In many lines, the POSelS advantageously replace more expensive systems composed of robots with grippers piloted by visual inspection systems to identify the individual orientations of the parts. This is especially true when the parts are small and the required fluxes are very high. From now on, we shall restrict our attention to the PODs.

Often several different PODs are available and only a fixed number of them are to be selected to build a POS. The basic principle is typically sketched in Fig. 3.2. It relies on the intuitive property that a loaded die which when dropped to

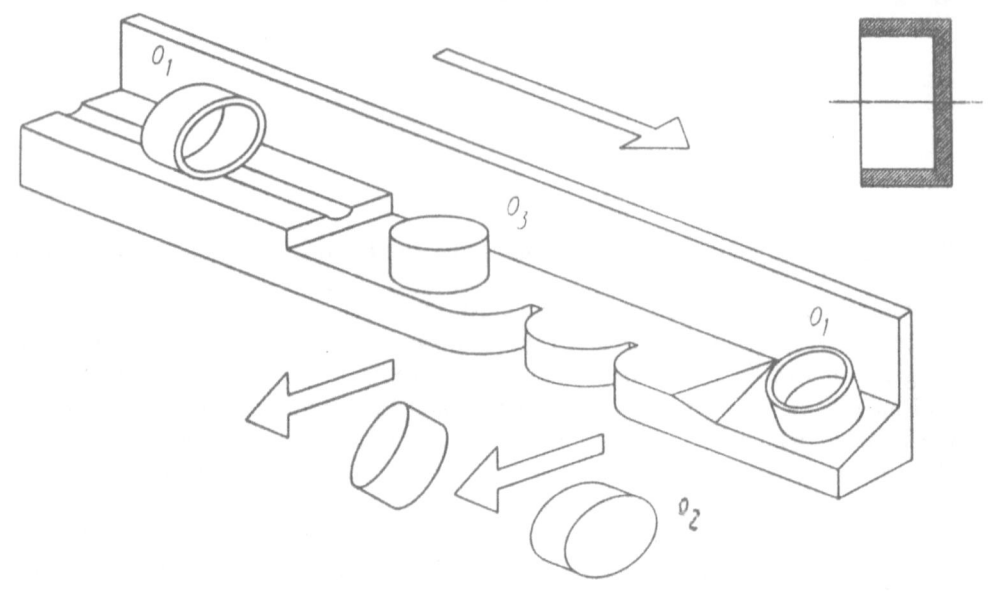

Fig. 3.1. Typical realization of a part orienting and selecting device. In this example, the parts can achieve four different orientations, O_1, O_2, O_3 and O_4. The step acts as a POD. It conserves the flux of incoming parts. The other devices are POSs.

the floor has a greater probability of presenting a particular side. To guarantee an even higher probability, one can think of throwing the dice at a large number of different heights. The natural question is therefore how to optimize the choice and the ordering of the selected PODs, (i.e. select the heights and the order in which the throwings are to be performed). This type of problem has been recently considered in the studies [SE], [JA1], [JA2]. The engineering realizations in this context are related in [MA1], [MA2], where programmable (i.e. flexible) POSelDs are discussed. A recent review of the general problem can be found in [YE].

Now we shall proceed by briefly reviewing the mathematical formalizations of a POS. We shall introduce a dynamical system viewpoint with a graphical representation that shall simplify and clarify the discussion.

3.2 The Device Selection and Ordering Problem. General Situation

Let us assume that a number N of not necessarily distinct POD are at our disposal to design a POS. The initial part distribution will be denoted by a probability vector, $\eta_{in} = (\eta_{in,1}, \eta_{in,2}, ..., \eta_{in,m})$; $\sum_{k=1}^{m} \eta_{in,k} \equiv 1$ with m being

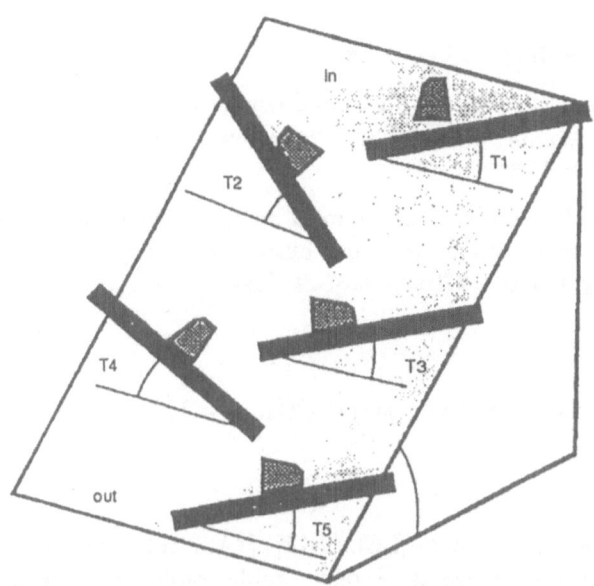

Fig. 3.2. Principle of a gravity part-feeder. The inclination angles $T1$ to $T5$ generate transfer matrices $\hat{\mathbf{T}}_1$ to $\hat{\mathbf{T}}_5$. This is a POS which clearly conserves the incoming flux of parts.

the number of possible orientations, (in Figure 3.1, for example $m = 4$). In the following, we shall assume, without loss of generality, that the probability of having the required orientation is always given by the first component of the probability vector $\boldsymbol{\eta}$, i.e. η_1.

The action of a single POD (say the device number k) on $\boldsymbol{\eta}_{in}$ is fully described by a transfer matrix $\hat{\mathbf{T}}_k$.

$$\boldsymbol{\eta}_{out} = (\eta_{out,1}, \eta_{out,2}, \ldots \eta_{out,m}) = \boldsymbol{\eta}_{in}\hat{\mathbf{T}}_k. \qquad (3.1)$$

In the following, we shall denote matrices by bold face capitals with a hat and vectors as greek bold face characters. The matrices $\hat{\mathbf{T}}_k$ are stochastic matrices (i.e. the sum of the elements of each line equals unity). This property explicitly expresses the conservation of the flux of parts through the POD.

The action of a POS composed of $n \leq N$ POD arranged in cascade can therefore be described by the product of n matrices $\hat{\mathbf{T}}_k$, $k = 1, 2, \ldots n$. We shall denote by $G(n)$ the set of all permutations of n POD and by $\psi(n) \in G(n)$, one particular permutation. The transfer matrix of a POS composed of n POD in the configuration $\psi(n)$ will be denoted by $\hat{\mathbf{T}}_{sys}[\psi(n)]$. This is itself a stochastic matrix (i.e. the product of stochastic matrices is a stochastic matrix, [MC], [ST]). Therefore we have:

$$\boldsymbol{\eta}_{out}[(\psi(n))] = (\eta_{out,1}[(\psi(n))], \eta_{out,2}[(\psi(n))], \ldots \eta_{out,m}[(\psi(n))]) =$$

$$= \eta_{in}\hat{\mathbf{T}}_{sys}[(\psi(n))] \qquad (3.2)$$

The most efficient arrangement is obviously achieved when the resulting POS produces the largest possible $\eta_{out,1}$. Accordingly, the general problem to be solved can now be re-stated as follows:

Problem: Select n, $(n \leq N)$ PODs and arrange them in an order such as to get the best possible efficiency. In other words, we must build a POS composed of n PODs in a configuration, say $\psi^*(n) \in G(n)$; $n \leq N$, leading to $\hat{\mathbf{T}}_{sys}[\psi^*(n)]$ with the property:

$$\eta_{out,1}[\psi^*(n)] \geq \eta_{out,1}[\psi(l)] \quad \forall \psi(l) \neq \psi^*(n) ; \ \psi(l) \in G(l), \quad l = 1, 2, ...N. \quad (3.3)$$

Various heuristic approaches proposed in [SE], [JA1], [JA2] are devoted to the solution of this problem which in general is of considerable complexity. There are two basic difficulties which are coupled, namely: the selection and the arrangement of the constituting POD's.

To illustrate the above concepts and approach to the ordering problem, it is natural to study the result obtained by commutating two devices $\hat{\mathbf{T}}_1$ and $\hat{\mathbf{T}}_2$. Let us then introduce an initial distribution $\eta_{in} = (\eta_{in,1}, \eta_{in,2})$ where $\eta_{in,2}$ represents the probability distribution vector of the $m-1$ states which are not in the required orientation. The matrices $\hat{\mathbf{T}}_k$; $k = 1, 2$ are written in the form:

$$\hat{\mathbf{T}}_k = \begin{pmatrix} a_k & \mathbf{b}_k \\ \mathbf{c}_k^* & \hat{\mathbf{D}}_k \end{pmatrix} \quad k = 1, 2, \qquad (3.4)$$

where $\mathbf{b}_k$ and $\mathbf{c}_k$ are $(l-1)$ dimensional row vectors, (the superscript $*$ denotes the operation of transposition) and $\hat{\mathbf{D}}_k$ is a $((l-1) \times (l-1))$ matrix. Now, we calculate the commutator $\Delta = \eta_{in}(\hat{\mathbf{T}}_1\hat{\mathbf{T}}_2 - \hat{\mathbf{T}}_2\hat{\mathbf{T}}_1)\mathbf{u}^*$ with $\mathbf{u} = (1, 0, ..., 0)$. A direct computation shows, [JA2]:

$$\Delta = \{\mathbf{b}_1\mathbf{c}_2^* - \mathbf{b}_2\mathbf{c}_1^*\} + \eta_{in,2}[\mathbf{b}_2^*\mathbf{c}_1\mathbf{Id} - \mathbf{c}_1\mathbf{b}_2^*]\mathbf{e}^* +$$

$$+ \eta_{in,2}[-\mathbf{b}_1\mathbf{c}_2^*\hat{\mathbf{Id}} + \mathbf{c}_2^*\mathbf{b}_1]\mathbf{e}^* + \eta_{in,2}[\hat{\mathbf{D}}_2\hat{\mathbf{D}}_1 - \hat{\mathbf{D}}_1\hat{\mathbf{D}}_2]\mathbf{e}^*, \qquad (3.5)$$

where $\mathbf{e}$ is a $(l-1)$ dimensional vector of the form $\mathbf{e} = (1, 1, ..., 1)$ and $\hat{\mathbf{Id}}$ stands for the identity matrix of dimension $(l-1) \times (l-1)$. Obviously, the result $\Delta > 0$ implies that the configuration $\hat{\mathbf{T}}_1 \hat{\mathbf{T}}_2$ is more efficient than the reverse one.

Eq. (3.5), explicitly shows that the ordering problem in general depends on the initial distribution η_{in}. However, when $m = 2$, Δ is independent of η_{in} (i.e. only the curly bracketed term in Eq. (3.5) survives) and then the problem is much more simple. This important case which is found to be instructive is discussed now.

3.3 Part Orienting System for Parts with Only Two Possible Orientations

Assume therefore that the parts which are feed into the POD can be found in one of only two possible orientations, their respective probabilities will be denoted by: $\eta_{in,1}$ for the correct (i.e. required) orientation and obviously $\eta_{in,2} = 1 - \eta_{in,1}$ for the wrong one. Hence, $\boldsymbol{\eta_{in}} = (\eta_{in,1}, 1 - \eta_{in,1})$. The POD acts as an operator on the initial probability distribution and should increase the probability of finding parts in the correct orientation. If $\boldsymbol{\eta_{out}} = (\eta_{out,1}, 1 - \eta_{out,1})$ denotes the probability distribution after the action of the POD, one expects $\eta_{out,1} \geq \eta_{in,1}$. For this simple situation, the action of the POD can be written as a 2×2 transfer matrix of the form:

$$\hat{\mathbf{T}} = \begin{pmatrix} (1-\alpha) & \alpha \\ \beta & (1-\beta) \end{pmatrix} \; ; \; 0 \leq \alpha, \beta \leq 1, \tag{3.6}$$

where $(1 - \alpha)$ denotes the probability that a part which enters the POD with the correct orientation stays in this correct orientation, α is the probability that a correctly oriented part entering the POD will leave it in the wrong orientation, β is the probability of entering in the wrong orientation and leaving with the correct one and finally $(1 - \beta)$ is the probability of entering and remaining in the wrong orientation after the action of the POD. The stochastic nature of the matrix Eq. (3.6) is explicitly written out. In the following, as no confusion is possible, we shall write:

$$\eta_{in,1} \equiv \eta_{in}. \tag{3.7}$$

In terms of the abbreviated notations η_{out} and η_{in}, we can write:

$$(\eta_{in}, (1 - \eta_{in}))\hat{\mathbf{T}} = (\eta_{out}, (1 - \eta_{out})). \tag{3.8}$$

Assume now that a POS is composed of n PODs. The action of the resulting device is:

$$\hat{\mathbf{T}}_{sys} = \prod_{k=1}^{n} \hat{\mathbf{T}}_k \tag{3.9}$$

and accordingly:

$$(\eta_{in}, (1 - \eta_{in}))\hat{\mathbf{T}}_{sys} = (\eta_{out}, (1 - \eta_{out})) \tag{3.10}$$

The product of matrices being non-commutative, one has to determine the order in which the succession of PODs have to be arranged to get the highest possible η_{out}. To discuss this question, we briefly review some properties of the product of stochastic matrices of the form Eq. (3.6).

Determinant.

$$D = \det \hat{\mathbf{T}} = 1 - (\alpha + \beta) \quad \text{and hence} \quad | D | \leq 1, \tag{3.11}$$

this follows from the fact that $\alpha \leq 1$ and $\beta \leq 1$ as they denote probability measures. Furthermore, we have:

$$| D_{sys} | = | \det \hat{\mathbf{T}}_{sys} | = | \prod_{k=1}^{N} \det \hat{\mathbf{T}}_k | = | \prod_{k=1}^{N} D_k | \leq 1, \qquad (3.12)$$

where we have introduced the notation: $D_k = \det \hat{\mathbf{T}}_k$

Spectral decomposition.

A property of 2×2 stochastic matrices is that one eigenvalue always equals unity, (i.e. $\lambda_1 = 1$). The other eigenvalue λ_2 therefore has a norm smaller than unity. We can always write:

$$\hat{\mathbf{T}} = \lambda_1 \hat{\mathbf{P}}_1 + \lambda_2 \mathbf{P}_2 = \mathbf{P}_1 + \lambda_2 \hat{\mathbf{P}}_2 \qquad (3.13)$$

with:

$$\lambda_2 = 1 - \alpha - \beta \qquad (3.14)$$

$$\hat{\mathbf{P}}_1 = \frac{1}{\alpha + \beta} \begin{pmatrix} \beta & \alpha \\ \beta & \alpha \end{pmatrix} \qquad (3.15)$$

and

$$\hat{\mathbf{P}}_2 = \frac{1}{\alpha + \beta} \begin{pmatrix} \alpha & -\alpha \\ -\beta & \beta \end{pmatrix} ; \qquad (3.16)$$

$\hat{\mathbf{P}}_1$ and $\hat{\mathbf{P}}_2$ are orthogonal projectors. This enables us to write immediately:

$$(\hat{\mathbf{T}})^n = \hat{\mathbf{P}}_1 + \lambda_2^n \hat{\mathbf{P}}_2. \qquad (3.17)$$

Graphical representation.

The action of $\hat{\mathbf{T}}$, can be represented graphically as an automorphism $\hat{\mathbf{T}} : I \rightarrow I$, where $I = [0, 1]$ is the unit interval. Indeed, as the initial probability vector $\eta = (\eta_{in}, 1 - \eta_{in})$ has a unit norm, the action of $\hat{\mathbf{T}}$ is completely characterised by its action on one of its components. We can straightforwardly express this by the following relation:

$$\eta_{out} = (1 - \alpha - \beta)\eta_{in} + \beta. \qquad (3.18)$$

By using Eq. (3.17), the combined action of a series of n identical PODs, can therefore be written as:

$$\eta_{out} = (1 - \alpha - \beta)^n [\eta_{in} - \frac{\beta}{\alpha + \beta}] + \frac{\beta}{\alpha + \beta}. \qquad (3.19)$$

This situation is sketched in Figure 3.3 when $D > 0$ and in Figure 3.4 in the case when $D < 0$.

As it is clear from Figures 3.2 and 3.3, the limit $\tilde{\eta}$ defined by:

$$\tilde{\eta} = \lim_{n \to \infty} \eta_{in} \hat{\mathbf{T}}^n = (\frac{\beta}{\alpha + \beta}, \frac{\alpha}{\alpha + \beta}) = (\tilde{\eta}, 1 - \tilde{\eta}) \ \forall \ \eta_{in} \in [0, 1] \ \Rightarrow$$

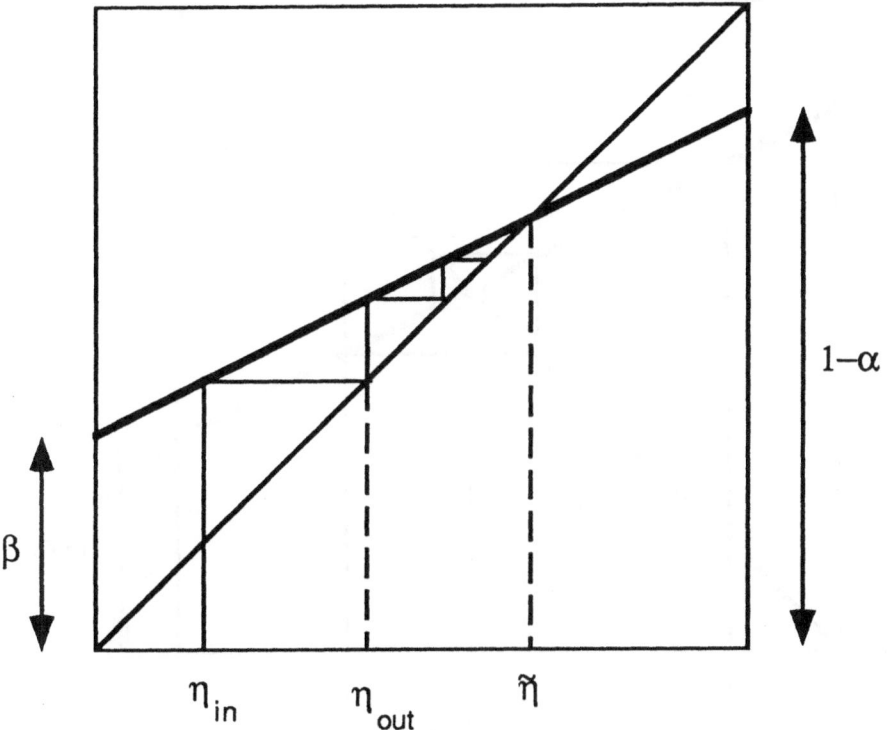

Fig. 3.3. Graphical representation of the action of a POD when the determinant of its transfer matrix is positive. η_{out} is the resuting distribution obtained after the action of one device and $\tilde{\eta}$ is the fixed point (attractor), which results from an infinite number of applications of identical devices.

$$\tilde{\eta} = \frac{\beta}{\alpha + \beta}. \tag{3.20}$$

$\tilde{\eta}$ corresponds to the abcissa of the point of intersection of the line $\mathcal{L}$ defined by $y = Dx + \beta$ with the bisectrix $y = x$. It is reached only after an infinite number of applications of similar PODs. The limit $\tilde{\eta}$ is reached monotonically if $D > 0$ and with an alternating sequence in the other case, see Figures 3.3 and 3.4. The graphic representations Figures 3.3 and 3.4 are those commonly adopted to describe the evolution of simple dynamical systems (see [LA] for instance and the numerous references therein). In dynamical systems jargon, $\tilde{\eta}$ is the attractor (here a fixed point), which is reached only after an infinite time (i.e. infinitely many iterations of the mapping $\hat{T}$). The slope D describes the dissipation rate of energy of an abstract dynamical system in which $I = [0, 1]$ plays the role of a section of the phase space, (i.e. the Poincaré section). When $D = 1$ the dynamical system is conservative (the length of the intervals are conserved, a property which, in the Poincaré section of the phase space, comes from the underlying Liouville theorem which holds for Hamiltonian dynamical

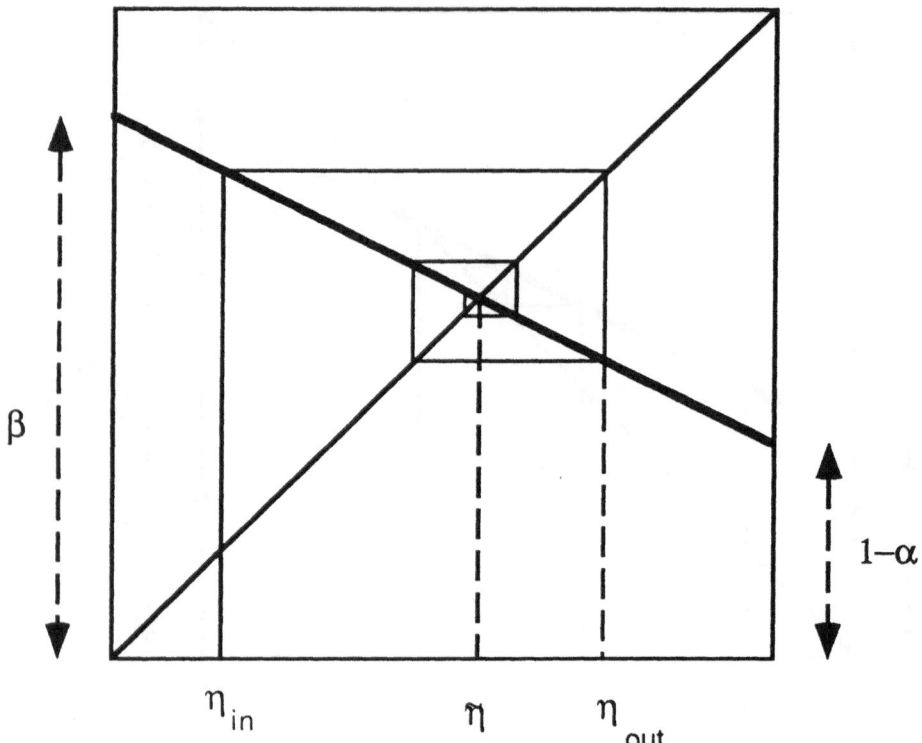

Fig. 3.4. Graphical representation of the action of a POD when the determinant of its transfer matrix is negative.

systems). The stability of the attractor (i.e. the stability of the fixed point $\bar{\eta}$) is clearly destroyed in the conservative case. The dissipation rate $D < 1$ of the dynamics is called the shrinkage rate in [JA2].

Consider now the POS obtained by the successive applications of n different PODs. This yields:

$$D_{sys} = \lambda_{sys} = \prod_{k=1}^{n}(1 - \alpha_k - \beta_k) = (1 - \alpha_{sys} - \beta_{sys}). \qquad (3.21)$$

The determinant D_{sys} given by Eq. (3.21) is independent of the order in which the PODs are arranged and therefore only α_{sys} and β_{sys} depend separately on the order but not their sum. A direct calculation shows that:

$$\beta_{sys} = \beta_1 D_2 D_3...D_n + \beta_2 D_3 D_4...D_n + ... + \beta_{n-1}D_n + \beta_n \qquad (3.22)$$

and similarly:

$$\alpha_{sys} = \alpha_1 D_2 D_3...D_n + \alpha_2 D_3 D_4...D_n + ... + \alpha_{n-1}D_n + \alpha_n \qquad (3.23)$$

and therefore, we obtain:

$$\tilde{\eta}_{sys} = \frac{\beta_{sys}}{\alpha_{sys} + \beta_{sys}} = \frac{\beta_{sys}}{1 - \lambda_{sys}}. \tag{3.24}$$

It is then clear from the graphs that changing the order of the POD units composing the POS only affects the ordinate at the origin of the line $\mathcal{L}_{sys}$ (i.e. β_{sys}) but not the slope λ_{sys}. Therefore, the most efficient device arrangement will be obtained for the arrangement leading to the largest possible β_{sys}. To achieve this we must distinguish between the following two cases:

a) $\underline{D_k = \lambda_k > 0 \ \forall \ k.}$

Recall that a permutation of n PODs is written by $\psi(n) \in G(n)$, where $G(n)$ is the set of all permutations on n PODs. The corresponding ordinate in the graphical representation Figure 3.2 is therefore $\beta[\psi(n)]$. Hence, the best POS will be obtained for the configuration $\psi^*(n)$ defined by:

$$\beta[\psi^*(n)] = \max_{\psi(n) \in G(n)} \beta[\psi(n)]. \tag{3.25}$$

The solution to the above problem is given by the following theorems [SE]:

Theorem 1 (ordering procedure).
The optimal configuration $\psi^*(n)$ for a POS composed of n PODs is obtained when

$$\frac{\beta_1}{\alpha_1} \le \frac{\beta_2}{\alpha_2} \le ... \le \frac{\beta_n}{\alpha_n}. \tag{3.26}$$

or equivalently:

$$\frac{\beta_1}{\alpha_1 + \beta_1} \le \frac{\beta_2}{\alpha_2 + \beta_2} \le .... \le \frac{\beta_n}{\alpha_n + \beta_n}. \tag{3.27}$$

The arrangement given by Eq. (3.27) leads to the highest possible β_{sys}. This can be seen directly from the expansion in Eq. (3.22). Observe in particular that Eq. (3.26) implies explicitly that $\Delta \ge 0$ in Eq. (3.5).

Theorem 2 (selection procedure).
Given an initial distribution η_{in}, the ordering of the PODs is given by theorem 1 and the selection consists of removing all devices for which we have: $\eta_{in} \ge \tilde{\eta} = \frac{\beta}{\alpha+\beta}$. After this operation the resulting configuration is called $\psi_0^*(n)$.

Looking at Fig. 3.3, it is also evident that the action of a POD when $\eta_{in} \ge \tilde{\eta}$ is to decrease the probability of finding the parts in the correct orientation. Therefore, these devices must indeed be removed.

Theorem 3 (selection procedure).
Suppose that in $\psi_0(n)$, there is one device say $\hat{T}_r$ for which $1 - \alpha = \beta \Rightarrow$ the action of T_r produces a vanishing slope of the line $\mathcal{L}_{sys}$. Then discarding all the devices that precede $\hat{T}_r$ in $\psi_0(n)$ leaves the efficiency of the POS unchanged.

Of course in this case the slope of $\hat{T}_{sys}$ is itself zero and from the expression Eq. (3.22) it is obvious that all terms containing the factor D_r do not contribute to β_{sys}.

b) Arbitrary signs of D_k

Here, the situation becomes more intricate. If the ordering problem alone is considered, Eq. (3.22) implies that the highest β_{sys} will result from the following ordering procedure:

a) Calculate the $\tilde{\eta}_k = \frac{\beta_k}{\alpha_k + \beta_k}$ and the corresponding determinant D_k, $\forall\, k = 1, 2, ...n$.

b) Define two ordering operators $\mathcal{O}^+$ and $\mathcal{O}^-$. The operator $\mathcal{O}^+$ arranges the PODs by increasing order of their $\tilde{\eta}_k$, (i.e. the largest $\tilde{\eta}_k$ being placed in the rightmost position). The action of $\mathcal{O}^-$ is opposite.

c) Place at the rightmost position the POD with the largest $\tilde{\eta}$; say it is $\hat{\mathbf{T}}_{i_1}$, then to the $n - 1$ devices which are to the left apply:

$$\begin{cases} \mathcal{O}^+, & \text{if } D_{i_1} = \det \hat{\mathbf{T}}_{i_1} > 0, \\ \mathcal{O}^-, & \text{if } D_{i_1} = \det \hat{\mathbf{T}}_{i_1} < 0. \end{cases} \qquad (3.28)$$

d) The second rightmost device is now given by $\hat{\mathbf{T}}_{i_2}$ Calculate the product $D_{i_1, i_2} = D_{i_1} D_{i_2}$ then to the $n - 2$ devices which are left, apply:

$$\begin{cases} \mathcal{O}^+, & \text{if } D_{i_1, i_2} > 0, \\ \mathcal{O}^-, & \text{if } D_{i_1, i_2} < 0. \end{cases}$$

e) Iterate until no PODs are left.

Clearly the above procedure is equivalent to the result obtained in Eq. (3.27) which is valid when $D_k > 0$, $\forall\, k$, (only the operator $\mathcal{O}^+$ is needed in this case). As an illustration consider six PODs with mixed signs D_k. Assume that we have:

$$| D_1^+ | \leq | D_2^+ | \leq | D_3^- | \leq | D_4^+ | \leq | D_5^- | \leq | D_6^- |, \qquad (3.29)$$

where the superscripts $+, -$ indicate the sign of the determinant. In this case the above ordering procedure produces the arrangement:

$$\psi^*(6) = 4, 5, 3, 2, 1, 6 \;\Rightarrow\; \hat{\mathbf{T}}_{sys} = \hat{\mathbf{T}}_4 \hat{\mathbf{T}}_5 \hat{\mathbf{T}}_3 \hat{\mathbf{T}}_2 \hat{\mathbf{T}}_1 \hat{\mathbf{T}}_6. \qquad (3.30)$$

When $m \geq 2$, (i.e. more than two orientations are possible), the problem is very complex. In particular, we have already observed in Eq. (3.5), that the solution depends on the initial probability vector η_{in}. Due to the stochastic nature of the transition matrices $\hat{\mathbf{T}}$, the general problem finds a natural mathematical framework in the study of finite inhomogeneous Markov chains. In this context, most of the results devoted to products of matrices concern their ergodicity properties and are therefore asymptotic in nature, (i.e. products of infinite series of stochastic matrices), [ST]. This is not of direct use here as we only deal with the arrangement of a relatively small number of PODs. A more directly relevant field of investigations can be found in [PA], where non-homogeneous Markov systems are discussed in connection with probabilistic automata. In this field, one indeed considers the properties of all possible products of stochastic matrices which can be formed from a fixed and finite set. We should also mention the exciting area of modern mathematical research dealing with the problem of convergence rates for Markov chains [DI], [AL] and the introductory paper by [RO].

The simplest approximation which comes to the mind is to group all the wrong orientation states into one single state. This leads to the approximation of the the action of the $l \times l$ transition matrix by a 2×2 matrix [SE]. This approximation scheme is well known in the framework of Markov chains. The coarse graining procedure which consists of grouping $(l-1)$ states into a single macro-state and to study the resulting approximated 2-state Markov chain is known as the lumpability procedure, [KS]. In particular, the 2-state approximation is exact if and only if the associated l-state Markov chain is lumpable into a binary state (i.e. 2 states), one of which represents the parts in the required orientation and the other the remaining ones [KS]. An even more refined concept known as weak lumpability [KS], [RS] which takes into account the initial probability vector η, might also prove useful and deserves attention for future research investigations; remember that the optimal ordering of devices depends in general on the initial distribution of components, see Eq. (3.5).

3.4 Summary and Perspectives

The overall assembly process is costly and difficult because the process seeks to impose organization and certainty on originally disorganized components. Parts that are provided in bulk lack any information about their position and orientation. Hence the difficulties and costs of automated assembly are due to this information gathering process. To achieve the reduction of configurational entropy, several part-orientors can be used namely: vibration, magnets, vacuum, air jets, wipers, ramps or edge risers, machine vision systems and robot/pick and place devices. The order in which these devices are to be arranged is a important issue for optimization studies which can be formalized by finite products of stochastic transition matrices and hence, the study of inhomogeneous Markov chains enter naturally into this problem. We can interpret the action of successive PODs on an initial distribution as the evolution of a dynamical system. Adopting this physical viewpoint, the fact that only finite products of matrices enter into the description indicates that rather than the equilibrium properties of the dynamics, it is the transient behaviour that is relevant in this context. The rich and complex behaviours inherent to the approach towards the attractors in dynamical systems enable one to capture the degree of complexity required in the optimization problem of part orienting mechanisms. Finally, it is worth pointing out that this optimization problem presents closed analogy with diagnostic of system failures [CW], [GI]. Assume indeed that we dispose of a series of tests to detect particular failures in a complex system. The result of a test does not fully characterize the state of the system (i.e. failed or good) it only increases the knowledge that we have about it. In this context, it is the optimal order and selection problem of available tests which has to be determined in order to detect the failure of a system with the highest possible probability.

4 Random Insertion Mechanisms

4.1 High Precision Automatic Insertion

At the level 0 of the general architecture of a production line (see Figure 1.1), a delicate problem in production line design concerns the flexible automation of very accurate assembly operations. While robotization offers the possibility of solving several practical problems, it often remains to overcome tedious difficulties such as the insufficient accuracy and repeatability in the performance of the manipulators, the inaccuracies in the gripping and the presentation of pieces, part-tolerances, etc. In particular, it often happens that the precision required to mate the parts is far below the robot's repeatability, especially when precise and chamferless insertions are considered. By repeatability, we mean the ability of a given manipulator to be cyclically commanded to attain an arbitrary fixed target. Among the solutions proposed, let us mention the *vibratory method*. This general philosophy consists in vibrating one of the parts in order to absorb relative misalignments. The vibrations enable the work tool to systematically explore a planar neighbourhood of fixed size in which the position errors between parts are confined. The approximate alignment, which enables the mating to be performed, is achieved when the distance between the parts is less than the characteristic clearance. This general problem has been recently discussed in the following references: [HF], [WA], [JE], [IV], [BD1], [BD2], [SC].

Restricting our attention to the configuration of Figure 4.1, we now consider a fixed circular hole and a peg which is activated until the insertion is performed (by reciprocity, the arrangement in which the hole is activated and the peg is fixed is identical). Let us denote by x_0 and y_0 the initial position of the peg relative to the hole, the position of whose centre whose coincides with the origin of the co-ordinates. The insertion will be possible at a time say t^* when the position $(x(t^*), y(t^*))$ of the peg satisfies the condition: $(x(t^*)^2 + y(t^*)^2) < a$, where a is the radius of a circle called the mating circle $\partial \mathcal{C}_m$ concentric with the hole disk (see Figure 4.2). The parameter a is called the clearance radius.

As the alignment process is time consuming, the basic problem is to explore, as efficiently as possible, a planar region in order to reach, as soon as possible, the positions where the mating is possible. Basically the choice between two alternative strategies has to be made. In one case, one deterministically sweeps the vicinity of the mating region with a chosen trajectory, typically a spiral, a Lissajous figure, etc..., [SC]. This solution is sketched in Figure 4.2 when a spiraling approach is adopted.

The mesh of the spiral m is chosen to be of the order a with $m \leq a$, (i.e. the difference between the diameters of the hole and the peg). Hence starting at

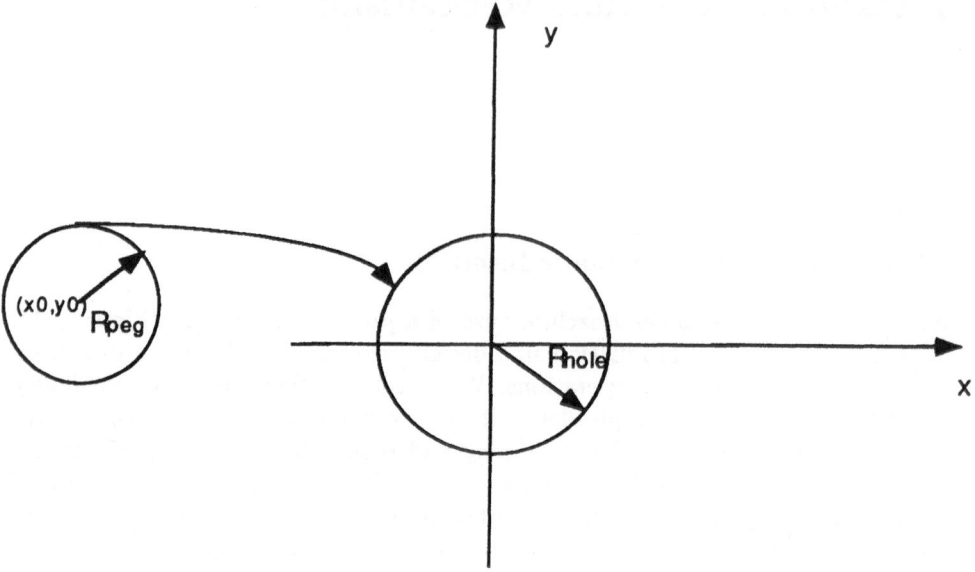

Fig. 4.1. Schematic view of the Peg-in-the-Hole Problem.

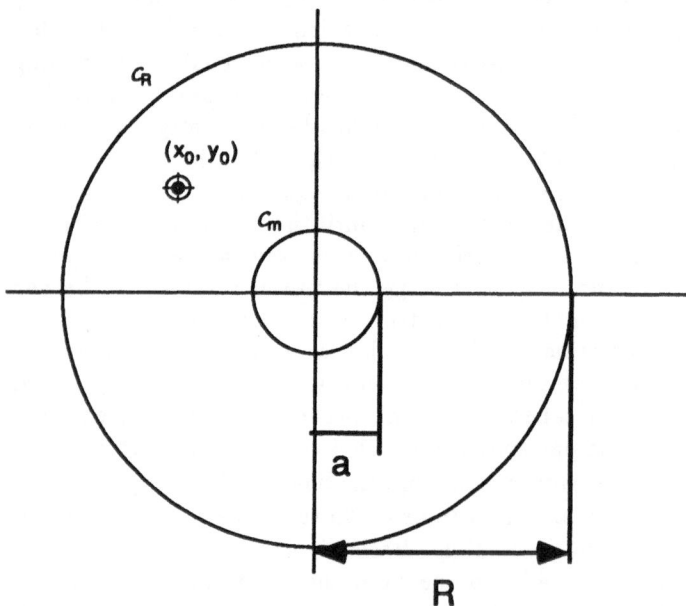

Fig. 4.2. Basic geometrical quantities used to describe the insertion problem. The exploration domain $\mathcal{G}$ from which the initial conditions (x_0, y_0) are drawn at random, is here chosen to be the disk $\mathcal{C}_R$ and the boundary of the disk $\mathcal{C}_m$ is the mating circle $\partial \mathcal{C}_m$.

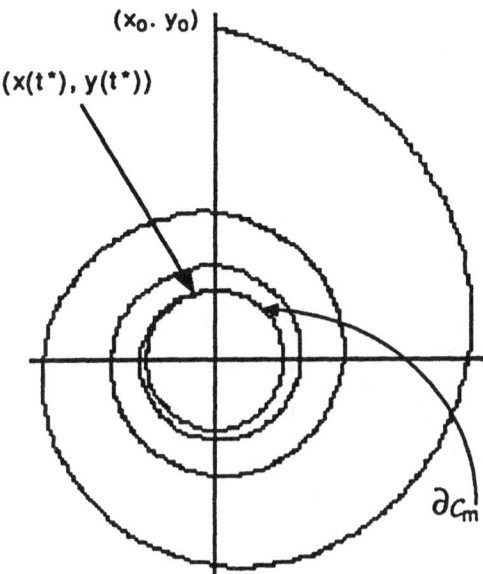

$(x_0 \cdot y_0)$

$(x(t^*), y(t^*))$

∂c_m

Fig. 4.3. Deterministic mating strategy achieved by a spiral motion.

a distance x from the origin, the mean insertion time T_{det} can be estimated as:

$$T_{det} = \frac{x-a}{a} \times \frac{1}{\Omega} \propto \frac{x}{a} \quad \text{when} \quad a \to 0 \qquad (4.1)$$

where Ω is the angular velocity of the spiraling motion which is assumed here to be constant and the mesh of the spiral has been taken as $m = a$ for this rough estimate. A similar estimation can be found in [SC] for the Lissajous figure approaches; the asymptotic behaviour for $a \to 0$ coincides with Eq. (4.1).

Another recent approach which is proposed is the random exploration of the vicinity of the mating region, [BD1], [BD2]. This strategy is sketched in Figure 4.4. Here, a purely random sweeping motion is induced and finally the mating region is reached.

What strategy should be used for small a? The deterministic approach or its random motion counterpart. This chapter is devoted to the study of stochastic exploration.

Important to our consideration is the fact that in a industrial environment, which is ultimately the one to be considered, external noise sources are ubiquitous and hence purely deterministic trajectories strictly speaking do not exist. There might be an even more primary reason to investigate the random search as a fundamentally efficient strategy. This can be traced back to the dynamical systems theory [AR], [GU], [LI], [MR], [LA]. Whilst a deterministic exploration strategy is ergodic, a random search tends to present a mixing behaviour [PE], [LI] which implies ergodicity. When mixing is present, the autocorrelations of initially close trajectories decay very fast. This property indicates that the search

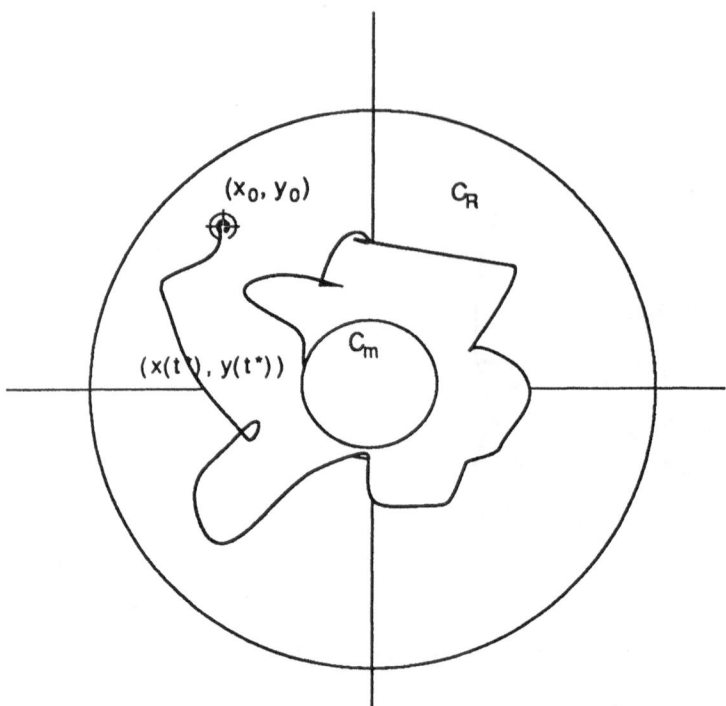

Fig. 4.4. Sketch of the random exploration strategy.

area will be rapidly and isotropically explored which results in potentially low mean mating time. An extended discussion concerning the hierarchy of chaotic behaviours and the relation between dynamical systems and probability theory has recently been given in [LA] and [JN]. In a very general context, it has been pointed out by [RU], that the random search philosophy may indeed often lead to surprisingly efficient results.

To try to quantify this heuristic reasoning, let us now discuss the random search approach when the excitation is idealized mathematically by a Gaussian White Noise (GWN) process.

4.2 Brownian and Related Motions in the Plane

We shall focus our attention on the peg-in-the-hole problem. The geometry we shall deal with is sketched in Figure 4.2. The clearance radius a is directly related to the precision required for the insertion operation. The initial positions are such that $(x_0, y_0) \in \mathcal{G}$, where $\mathcal{G}$ is an exploration domain in the neighbourhood of the origin. The size and shape of $\mathcal{G}$ depend on the geometric and reproducibility characteristics of the robot or manipulator involved. When the insertion trials are repeated, the (x_0, y_0) are drawn at random from $\mathcal{G}$. Here, we shall confine our attention to the case where $\mathcal{G}$ is a disk of radius R denoted by $\mathcal{C}_R$. The problem is now to select a dynamical evolution governing the co-ordinates of the peg (x_t, y_t) in order to have $(x(t^*)^2 + y(t^*)^2) < a$ with the smallest possible t^*. The random approach strategy which will be discussed now is appealing for two main reasons:

a) The noise is continuously and unavoidably present in industrial environments. Therefore, whatever the dynamics used, external fluctuations will always alter the dynamics.

b) The random approach presents up to a certain extent a flexibility of use. Indeed, it can be implemented without modification in situations with different mating precision radii.

We start by considering the simplest random dynamics for the motion of a peg. This evolution is given by the following Stochastic Differential Equations (SDE):

$$dx(t) = \sigma_x dW_{x,t} \tag{4.2}$$

and

$$dy(t) = \sigma_y dW_{y,t}. \tag{4.3}$$

where σ_x and σ_y are diffusion constants with physical units $[\frac{m}{s^{\frac{1}{2}}}]$, $dW_{x,t}$ and $dW_{y,t}$ are two independent GWN, hence we have [KA], [GA], [HS], [RI]:

$$\langle dW_{x,t} \rangle = \langle dW_{y,t} \rangle = 0, \tag{4.4}$$

$$\langle dW_{x,t}\ dW_{x,\tau} \rangle = \langle dW_{y,t}\ dW_{y,\tau} \rangle = \delta(|\,t - \tau\,|) \tag{4.5}$$

and from the independence of the noise sources, the cross-correlations are:

$$\langle dW_{x,t}\ dW_{y,\tau} \rangle = 0, \quad \forall\, t, \tau. \tag{4.6}$$

The stochastic differential equations (SDE), Eqs. (4.2) and (4.3) can be transformed into polar co-ordinates (r, θ). In the isotropic case, i.e. when $\sigma_x = \sigma_y = \sigma$, this yields, [GA]:

$$dr = \left(\frac{\sigma^2}{2r} \right) dt + \sigma dW_{1,t} \tag{4.7}$$

and

$$d\theta = \left(\frac{\sigma}{r} \right) dW_{2,t} \tag{4.8}$$

with $dW_{1,t}$ and $dW_{2,t}$ being once more independent GWN processes. While in actual configurations, it often happens that circular symmetry approximately holds, the radial motion may nevertheless exhibit a systematic drift. Furthermore, the diffusion coefficients may also differ along the radial and the angular directions. To cope with these more general situations, Eqs. (4.7) and (4.8) will be generalized by the following SDE:

$$dr = \left(-\delta(r) + \frac{\sigma_2^2}{2r}\right) dt + \sigma_1 dW_{1,t} \qquad (4.9)$$

and

$$d\theta = \left(\frac{\sigma_2}{r}\right) dW_{2,t}. \qquad (4.10)$$

In Eq. (4.9), the drift $\delta(r) \geq 0$, represents a net tendency, (i.e. heading velocity), to approach the origin of the co-ordinates and σ_1, σ_1 are diffusion constants which differ along the radial and angular directions. Hence, when $\delta(r) > 0$, Eqs. (4.9) and (4.10) describe a pole-seeking Brownian Motion (BM). Again, the radial and angular noises are assumed to be independent GWN. For $\delta(r) = \delta = $ const, the model described in Eqs. (4.9) and (4.10) were introduced in [KD].

Remember that we have assumed that the exploration domain is the disk $\mathcal{C}_R$. Hence, the behaviour of the dynamics has to be specified when the motion reaches the boundary circle $\partial \mathcal{C}_R$.

For the present application, only the radial part of the motion will be relevant. Indeed, we are only interested in the time required to touch the mating circle $\partial \mathcal{C}_m$ but not in the angular distribution of the impacts on $\partial \mathcal{C}_m$. Note however that the angular impact probability distribution is explicitly derived in [KD], [ME], [PI1], [PI2]. This information is relevant if, for the dynamics described by Eqs. (4.9) and (4.10), the impacts on $\partial \mathcal{C}_m$ are not uniformly distributed. A simple symmetry argument enables one to conclude that in this case the initial positions x_0, y_0 are themselves not uniformly distributed in the disk $\mathcal{C}_R$. Therefore, by judiciously translating the position of the hole and by decreasing the size of $\mathcal{C}_R$, until uniformity of the impacts on $\partial \mathcal{C}_m$ is achieved, one is able to potentially diminish the exploration time .

With the use of the GWN processes, the dynamics is a Markovian diffusion process [GA], [KA]. The complete probabilistic information is therefore contained in the Transition Probability Density (TPD) denoted by: $P(r, \theta, t \mid r_0, \theta_0, t_0 = 0)$ which obeys the Fokker-Planck Equation (FPE), [RI], [GA], [KA], [VK]:

$$\frac{\partial}{\partial t} P(r, \theta, t \mid r_0, \theta_0, t_0 = 0) = (\mathcal{F}_r + \mathcal{F}_\theta) P(r, \theta, t \mid r_0, \theta_0, t_0 = 0), \qquad (4.11)$$

where the second order differential operators corresponding to the SDE given by Eqs. (4.9) and (4.10) have the forms:

$$\mathcal{F}_r(.) = -\frac{\partial}{\partial r}\left(-\delta(r) + \frac{\sigma_2^2}{2r}\right)(.) + \frac{\sigma_1^2}{2}\frac{\partial^2}{\partial r^2}(.) \qquad (4.12)$$

and

$$\mathcal{F}_\theta(.) = \frac{\sigma_2^2}{2r}\frac{\partial^2}{\partial \theta^2}(.) . \qquad (4.13)$$

Due to the rotational symmetry, the radial diffusion can be treated independently of the angular one. Without lost of generality, we can take the initial angle as $\theta_0 = 0$. Hence, the radial (marginal) TPD $P_M(r,t \mid r_0, 0)$:

$$P_M(r,t \mid r_0, 0) = \int_0^{2\pi} P(r,\hat{\theta},t \mid r_0, \theta_0 = 0, t_0 = 0)d\hat{\theta} \qquad (4.14)$$

will be a solution of the FPE:

$$\frac{\partial}{\partial t} P_M(r,t \mid r_0, 0) = \mathcal{F}_r P_M(r,t \mid r_0, 0). \qquad (4.15)$$

By separation of the space and time variables, we can rewrite the solution of Eq. (4.15) in the form, [GA], [KA], [RI], [VK]:

$$P_M(r,t \mid r_0, 0) = \sum_{\{\lambda_n\}} \Pi_{\lambda_n} \exp\{-\lambda_n \sigma_1^2 t\} \phi_n(r) \phi_n(r_0) P_{M,S}(r), \qquad (4.16)$$

where $P_{M,S}(r)$ stands for the stationary (i.e. time independent) probability density, $\phi_n(r)$ and $\{\lambda_n\}$ respectively are the eigenfunctions and the spectrum of the associated Sturm-Liouville problem, on the domain defined by $r \in \mathcal{C}_R$ and $\theta \in [0, 2\pi]$, [GA], [KA], [RI], [VK]. The coefficients Π_{λ_n} are determined by the initial TPD, i.e. $P(r,t = 0 \mid r_0, 0) = \delta(r - r_0)$ which leads to [KA]:

$$\frac{1}{\Pi_{\lambda_n}} = \int_{\Omega} \phi_n^2 P_{M,S}(r)dr. \qquad (4.17)$$

This general framework is now illustrated in two explicit situations, namely:

i) $\delta(r) = 0$ and a reflecting behaviour on the circle $\partial \mathcal{C}_R$

For this case we have:

$$r^{\nu} \frac{\partial}{\partial r} \Phi_{\lambda}(r) \mid_{r=0+} = 0 \qquad (4.18)$$

with the definition:

$$\nu = \frac{\sigma_2^2}{\sigma_1^2}. \qquad (4.19)$$

The other boundary behaviour which expresses the reflecting behaviour on the circle $\partial \mathcal{C}_R$ has the form, [KA]:

$$\frac{\partial}{\partial r} \Phi_{\lambda}(r) \mid_{r=R} = 0. \qquad (4.20)$$

In this case, the expansion Eq. (4.12) leads to :

$$P(r,t \mid r_0, 0) = (\frac{r}{r_0})^{\frac{\nu}{2}} \sqrt{rr_0} \sum_{n=0}^{\infty} \Pi_{\lambda_n} \exp\{-\frac{j_{\frac{\nu+1}{2},n}^2 \sigma_1^2}{2R^2}t\} J_{\frac{\nu-1}{2}}(\frac{j_{\frac{\nu+1}{2},n} r}{R}) J_{\frac{\nu-1}{2}}(\frac{j_{\frac{\nu+1}{2},n} r_0}{R})$$

$$\qquad (4.21)$$

where $j_{\frac{\nu+1}{2},n}$, $n = 1,2,3...$ are the zeros of the Bessel functions $J_{\frac{\nu+1}{2}}(z)$, [GR]. The $j_{\frac{\nu+1}{2},n}$, $n = 1,2,3...$ constitute the spectrum of the Sturm-Liouville problem. We omit an explicit calculation of the coefficients Π_{λ_n} as we shall not use them directly. When the radius of $\mathcal{C}_R$ becomes infinite, i.e. when $R \to \infty$, Eq. (4.20) can be summed-up in the well known compact form, [KA]:

$$P(r,t \mid r_0, 0) = \frac{r^\nu}{\sigma^2 t}(rr_0)\exp\{-\frac{r^2 + r_0^2}{2\sigma^2 t}\}I_{\frac{\nu-1}{2}}(\frac{rr_0}{\sigma^2 t}) \qquad (4.22)$$

with $I_{\frac{\nu-1}{2}}(\frac{rr_0}{\sigma^2 t})$ a modified Bessel function [GR]. The process characterized by Eq. (4.22) is known as the Bessel diffusion process. When $\nu = N - 1$ with $N = 1,2,3,...$, this process describes the Brownian motion in dimension N. A wealth of theorems describing the behaviour of this process is available. Without trying to be exhaustive, let us mention [CI], [IT], [KA] where results regarding the sojourn time in a ball $\mathcal{B}$ and the exit time out of this ball are discussed.

ii) $\delta r = \alpha r$ and $R = \infty$

In this case, the diffusion process is known as the radial Ornstein-Uhlenbeck process [KA]. The TPD in this case reads, [KA]:

$$P(r,t \mid r_0, 0) = 2(\frac{\alpha}{\sigma_1^2})^{\beta+\frac{1}{2}}r^{2\beta+1}e^{-\frac{\alpha r^2}{\sigma_1^2}}\frac{1}{1 - e^{-2\alpha t}} \times$$

$$\times \exp\{\frac{\frac{-\alpha}{\sigma_1^2}(r^2 + r_0^2)e^{-2\alpha t}}{1 - e^{-2\alpha t}}\}\chi^{-\beta}I_\beta(\frac{2\chi}{1 - e^{-2\alpha t}}), \qquad (4.23)$$

with the definitions $\beta = \frac{\sigma_2^2}{2\sigma_1^2} - \frac{1}{2}$ and $\chi = \frac{\alpha r r_0 e^{-\alpha t}}{\sigma_1^2}$.

Having the transition probability density, it is, in principle, possible to obtain the first hitting time distribution of the process $r(t)$, given by Eq. (4.7), to a barrier. The complete solution to this problem is obtained in terms of Laplace transformations [CI], [KT]. For the application we have in mind, we are mostly interested in the mean first access time to $\partial\mathcal{C}_m$, i.e. the mean first hitting time $\langle T_a(r_0)\rangle$ of the process $r(t)$ at the radial position $r = a$ if $r(t = 0) = r_0 \geq a$ (see Figure 4.4). Due to the fact that the process is a diffusion, $\langle T_a(r_0)\rangle$ obeys a differential equation given by [KA]:

$$\frac{\sigma_1^2}{2}\frac{\partial^2}{\partial r_0^2}\langle T_a(r_0)\rangle + (-\delta(r_0) + \frac{\sigma_2^2}{2r_0})\frac{\partial}{\partial r_0}\langle T_a(r_0)\rangle = -1. \qquad (4.24)$$

The boundary condition at $r_0 = a$ obviously reads:

$$\langle T_a(r_0 = a)\rangle = 0, \qquad (4.25)$$

while $\langle T_a(r_0)\rangle$ with $r_0 \in \partial\mathcal{C}_R$ is specified by the boundary behaviour of the diffusion on the border of the exploration domain defined by the the circle $\partial\mathcal{C}_R$. In particular, for a reflecting condition on $\partial\mathcal{C}_R$, we must impose the condition: $\frac{\partial}{\partial r_0}\langle T_a(r_0)\rangle\mid_{r_0=R} = 0$.

Now, we integrate Eq. (4.24) in three different special cases:

a) $\delta(r) \equiv 0$ and a reflecting boundary at $r = R$.

In this case, the integration of Eq. (4.24) yields:

$$\langle T_a(r_0)\rangle = \begin{cases} \frac{R^2}{\sigma_1^2}\ln(\frac{r_0}{a}) - (\frac{r_0^2-a^2}{2\sigma_1^2}) & \text{when } \nu = 1 \\ \frac{2R^{\nu+1}}{\sigma_1^2(1-\nu^2)}(r_0^{2\nu} - a^{2\nu}) - \frac{r_0^2-a^2}{\sigma_1^2(1+\nu)} & \text{otherwise.} \end{cases} \tag{4.26}$$

b) $\delta = $ constant and a natural boundary at $r \to \infty$.

In this case, we obtain:

$$\langle T_a(r_0)\rangle = \frac{1}{\delta}(r_0 - a) + \frac{\sigma_1^2}{2\delta^2}\ln(\frac{r_0}{a}) + C(\delta, \nu, r_0, a), \tag{4.27}$$

with

$$C(\delta, \nu, r_0, a) = \sum_{k \geq 2}^{\infty}(\frac{\sigma_1^2}{2\delta})^k(\frac{1}{1-k})\nu(\nu - 1)..(\nu - k + 1)(r_0^{1-k} - a^{1-k}). \tag{4.28}$$

Remark that $C(\delta, \nu = 1, r_0, a) = 0$. The situation $\nu = 1$ is encountered in most applications. Indeed, it corresponds to the standard (isotropic) two-dimensional BM for which we have: $\sigma_1 = \sigma_2 (\Rightarrow \nu = \frac{\sigma_1^2}{\sigma_2^2} = 1)$. The result given by Eq. (4.27) in the case $\nu = 1$ has been derived in [KD] by using the complete distribution of the first hitting time.

c) $\delta r = \alpha r$, $\nu = 1$ and $R \to \infty$

Here, we have:

$$\langle T_a(r_0)\rangle = \frac{1}{\alpha}\ln(\frac{r_0}{a}). \tag{4.29}$$

Note that Eq. (4.29) does not involve the diffusion constant σ. This is accidental for the choice $\delta(r) = \alpha r$ and $\nu = 1$. In fact, the explicit occurrence of the diffusion parameter σ will enter into the expressions for the highest moments $\langle T_a^n(r_0)\rangle$, $n \geq 2$. Observe that the result Eq. (4.29) can be interpreted from a purely deterministic reasoning. Let us indeed consider the deterministic motion (i.e. $\sigma = 0$ in Eq. (4.9)). We therefore immediately have:

$$\frac{d}{dt}r(t) = -\alpha r(t) \Rightarrow r(t) = r_0 \exp(-\alpha t). \tag{4.30}$$

This equation describes the relaxation of $r(t)$, (with a rate α), toward the origin. The time needed to attain the position a if, at time $t = 0$, the motion started at $r_0 \geq a$ is therefore given by Eq. (4.29).

The higher order moments $\langle T_a^n(r_0)\rangle$ with $n \geq 2, 3, ...$ also possess an amount of informative content. These moments, $\langle T_a^n(r_0)\rangle$, can be calculated by solving a generalization of Eq. (4.24), [KA] which reads:

$$\frac{\sigma_1^2}{2}\frac{\partial^2}{\partial r_0^2}\langle T_a^n(r_0)\rangle + (-\delta(r_0) + \frac{\sigma_2^2}{2r_0})\frac{\partial}{\partial r_0}\langle T_a^n(r_0)\rangle = -n\langle T_a^{n-1}(r_0)\rangle \qquad (4.31)$$

for which $n = 1, 2, 3...$. Once more, we have $\langle T_a^n(r_0 = a)\rangle = 0$ and we must impose the relevant boundary conditions when $r_0 \in \partial C_R$.

Returning to probability distributions, the marginal stationary probability density $P_{M,S}(r)$ obtained in the limit $t \rightarrow \infty$, can easily be calculated as microscopic reversibility holds for this one-dimensional diffusion process, [GA]. For the above cases a), b) and c), we respectively obtain:

$$P_{M,S}(r) = \frac{2r}{R^2}, \qquad (4.32)$$

$$P_{M,S}(r) = \frac{1}{\Gamma(\nu+1)}(\frac{2\delta}{\sigma_1})^{\nu+1}r^\nu\exp\{-\frac{2\delta}{\sigma_1^2}r\} \qquad (4.33)$$

and finally

$$P_{M,S}(r) = \frac{\sigma^2}{2\alpha}\,r\exp\{-\frac{\alpha}{\sigma^2}r^2\}. \qquad (4.34)$$

4.3 Discussion of the Analytical Results in the Robotic Context

What have we learnt applicable to the robotic context from the aforegoing analysis?

a) From the case $\delta = 0$ with a reflecting boundary on the circle ∂C_R and for an isotropic diffusion (i.e. $\nu = 1$), the stationary probability measure $P_{M,S}(r)$ is given by Eq. (4.32). This corresponds to a uniform probability density on the disk C_R. The reflecting boundary condition expresses the fact that the exploratory motion is confined to the disk C_R and whenever the boundary ∂C_R is attained, the peg is simply reflected back to C_R. The relaxation toward the stationary probability measure $P_{M,S}(r)$ can be read from Eq. (4.21). The exponential time dependent factors under the summation introduce a characteristic diffusion time τ_{diff} which, asymptotically, governs the speed of approach to the equilibrium measure $P_{M,S}(r)$, ultimately reached at $t = \infty$. The value of τ_{diff} is given by the size of the gap between the zero and the lowest eigenvalue of the spectrum, i.e. $\lambda_{n=1}$. Due to the reflecting boundary imposed on the dynamics, a non-vanishing stationary measure $P_{M,S}(r)$ is attained. Therefore the eigenvalue $\lambda_{n=0} = 0$ belongs to the spectrum. Hence, we have:

$$\tau_{diff} \propto (\lambda_{n=1} - \lambda_{n=0}) \propto \frac{R^2}{\sigma_1^2}. \tag{4.35}$$

Remark that precisely the term $\frac{R^2}{\sigma_1^2}$ enters, as a prefactor, into the expression of the mean first access time $\langle T_a(r_0)\rangle$, which for this case is given by Eq. (4.26). It confirms our intuition that, the smaller the surface of the exploration zone is, the more rapidly the insertion is achieved. The contribution $\frac{1}{\sigma_1^2}$ in Eq. (4.35) shows the importance of having a large diffusion parameter.

We point out that the parameter a, which characterises the precision required to achieve the mating, only influences $\langle T_a(r_0)\rangle$ logarithmically. In the isotropic case $\nu = 1$, Eq. (4.26) reads:

$$\langle T_a(r_0)\rangle = \frac{R^2}{\sigma_1^2}\ln(\frac{r_0}{a}) \propto \tau_{diff}\ln(\frac{r_0}{a}). \tag{4.36}$$

This behaviour has to be contrasted with a deterministic spiralling approach for which the mean insertion time is estimated in Eq. (4.1). In this deterministic approach, $\langle T_a(r_0)\rangle$ is, when $a \to 0$, hyperbolic in the clearance parameter a.

The discussion of the situations when non-vanishing headings are present (i.e. cases b) and c)) introduces the concept of a confusion circle.

For the case $\delta = $ constant, [KD], the stationary probability density $P_{M,S}(r)$ is given by Eq. (4.33) and the first access time is given by Eq. (4.27). $\langle T_a(r_0)\rangle$ contains two contributions. The first contribution in the RHS of Eq. (4.27) simply expresses the time required for a direct access to the mating circle ∂C_m when the heading velocity is δ. The second contribution, which again exhibits a logarithmic behaviour, is entirely due to the fluctuations. This logarithmic contribution clearly dominates when $a \to 0$. The prefactor in front of the logarithmic term which we denote by $R_{conf} = \frac{\sigma_1^2}{2\delta}$ is the radius of the circle of confusion C_{conf} defined by: $\frac{\partial}{\partial r}P_{M,S}(r)\mid_{r=R_{conf}} = 0$. The origin of the name "circle of confusion", first introduced in [KD], comes from the fact that for initial conditions $r_0 > R_{conf}$, the mean tendency is to approach the mating circle ∂C_m and inversely if $r_0 < R_{conf}$, the motion winds around ∂C_m before this circle is touched and the insertion is performed. Note that provided $\delta > 0$, $\langle T_a(r_0)\rangle$ is finite even when the exploration disk C_R becomes arbitrarily large which is the situation covered by our calculation [KD].

When the heading is $\delta(r) = \alpha r$, the radial motion again defines a pole-seeking BM with a heading which increases linearly with the distance to the origin of the co-ordinates. This situation is precisely encountered in the actual realization to be described later. The origin of this effective heading is due in practice to the actuators and the noise generator [HO3]. For this case the circle of confusion ∂C_{conf} has a radius defined by $\frac{\partial}{\partial r}P_{M,S}(r)\mid_{r=R_{conf}} = 0$, where the stationary probability density $P_{M,S}(r)$ is given by Eq. (4.34). Hence we obtain $R_{conf} = $

$\sqrt{\frac{\sigma^2}{2\alpha}}$. As in the previous situations, we observe that $\langle T_a(r_0)\rangle \propto \ln(\frac{r_0}{a})$, see Eq. (4.29). Again the prefactor $\frac{1}{\alpha}$ in Eq. (4.29) coincides with the characteristic diffusion time τ_{diff} which can be read in the time dependent TPD given by Eq. (4.23).

The logarithmic behaviour of $\langle T_a(r_0)\rangle$ which is characteristic of the random search, has been observed by means of the experimental set up which will be presented in section 4.5.

4.4 Exploration with Coloured Noise

So far the analysis has been restricted to the GWN excitations. Now, we study the dynamics in the presence of coloured noise, (i.e. finitely correlated stochastic processes). For a general Gaussian noise with continuous realizations and a rational spectral density, it is always possible to describe the dynamics by constructing an ad-hoc set of SDE with GWN in higher dimensions (i.e. the dimension depends on the power of the denominator of the rational spectral density) [SV]. Here, we shall study the simplest case (i.e. when the spectral density is a Lorentzian). The exploration process in the plane is given by the stochastic differential equations (SDE):

$$dx = u_1(t)dt, \qquad (4.37)$$

and

$$dy = u_2(t)dt, \qquad (4.38)$$

where the processes $u_1(t)$ and $u_2(t)$ are finitely correlated and are here chosen as Ornstein-Uhlenbeck (OU) processes, themselves defined by:

$$du_1(t) = -\lambda u_1(t)dt + \sigma\lambda dW_{1,t}, \qquad (4.39)$$

and

$$du_2(t) = -\lambda u_2(t)dt + \sigma\lambda dW_{2,t}, \qquad (4.40)$$

where $dW_{1,t}$ and $dW_{2,t}$ are independent GWN.

The FPE governing the 4-dimensional degenerate diffusion defined by Eqs. (4.37) - (4.40), reads:

$$\frac{\partial}{\partial t}P(x,y,u_1,u_2,t \mid x_0,y_0,u_{1,0},u_{2,0},0) =$$

$$= (\mathcal{F}_x + \mathcal{F}_y)P(x,y,u_1,u_2,t \mid x_0,y_0,u_{1,0},u_{2,0},0), \qquad (4.41)$$

where the FP operators $\mathcal{F}_x$ and $\mathcal{F}_y$ respectively can be written as:

$$\mathcal{F}_x(.) = -u_1\frac{\partial}{\partial x}(.) + \frac{\partial}{\partial u_1}\lambda u_1(.) + \frac{\sigma^2\lambda^2}{2}\frac{\partial^2}{\partial u_1^2}(.) \qquad (4.42)$$

and

$$\mathcal{F}_y(.) = -u_2 \frac{\partial}{\partial y}(.) + \frac{\partial}{\partial u_2} \lambda u_2(.) + \frac{\sigma^2 \lambda^2}{2} \frac{\partial^2}{\partial u_2^2}(.) . \qquad (4.43)$$

Due to the fact that the x and y directions are not correlated, the TPD factorizes into two parts. Moreover, the initial conditions are:

$$P(x, y, u_1, u_2, t = 0 \mid x_0, y_0, u_{1,0}, u_{2,0}, 0) =$$

$$= \delta(x - x_0)\, \delta(y - y_0)\, \mathcal{N}_{u_1}(0, \frac{\sigma^2 \lambda}{2})\, \mathcal{N}_{u_2}(0, \frac{\sigma^2 \lambda}{2}), \qquad (4.44)$$

where $\mathcal{N}_z(m, v)$ stands for the normal density of the random variable z with mean m and variance v.

Let us now solve the FPE for the x-component of the process. To this end, we introduce the time dependent change of variables $(x, u_1) \rightarrow (I_1, I_2)$ defined by [CD]:

$$I_1 = x + \frac{u_1}{\lambda} \qquad (4.45)$$

and

$$I_2 = u_1 e^{\lambda t}. \qquad (4.46)$$

In terms of these new variables, the FPE is purely diffusive, namely:

$$\frac{\partial}{\partial t} P(I_1, I_2, t \mid I_{1,0}, I_{2,0}, 0) = \Delta(t) P(I_1, I_2, t \mid I_{1,0}, I_{2,0}, 0), \qquad (4.47)$$

where the time-dependent diffusion operator $\Delta(t)$ reads:

$$\Delta(t) = \frac{\sigma^2}{2} \frac{\partial^2}{\partial I_1^2} + \lambda \sigma^2 e^{\lambda t} \frac{\partial}{\partial I_1} \frac{\partial}{\partial I_2} + \frac{\lambda \sigma^2}{2} \frac{\partial^2}{\partial I_2^2}. \qquad (4.48)$$

The solution to Eq. (4.42) is immediately obtained in the form, [CD]:

$$P(I_1, I_2, t \mid I_{1,0}, I_{2,0}, 0) = \frac{1}{2\pi \sqrt{D(t)}} exp^{(-\gamma(I_1, I_2, t))}, \qquad (4.49)$$

with

$$\gamma(I_1, I_2, t) = -\frac{1}{2D(t)} [\alpha(t)(I_1 - I_{1,0})^2 +$$

$$+ 2\eta(t)(I_1 - I_{1,0})(I_2 - I_{2,0}) + \beta(t)(I_2 - I_{2,0})^2], \qquad (4.50)$$

where we have:

$$\alpha(t) = \frac{\lambda \sigma^2}{2} [e^{2\lambda t} + A] \qquad (4.51)$$

$$\beta(t) = \sigma^2 t + B \qquad (4.52)$$

$$\eta(t) = -\sigma^2 [e^{\lambda t} + C] \qquad (4.53)$$

and

$$D(t) = \alpha(t)\beta(t) - \eta^2(t). \qquad (4.54)$$

The constants A, B and C are determined by the initial conditions and read:

$$A = 0, \ B = \frac{\sigma^2}{2\lambda} \ \text{ and } \ C = -\frac{1}{2}. \qquad (4.55)$$

Using these expressions, it is tedious but elementary to express the radial part of the TPD. This reads:

$$P_M(r, t \mid r_0, 0) = \frac{r}{\tau(t)} e^{-\left(\frac{r^2 - r_0^2}{2\tau(t)}\right)} \mathcal{I}_0\left(\frac{rr_0}{\tau(t)}\right), \qquad (4.56)$$

with

$$\tau(t) = \sigma^2 t - \frac{\sigma^2}{\lambda}(1 - e^{-\lambda t}). \qquad (4.57)$$

For $\lambda \to \infty$, Eq. (4.57) is precisely identical to the radial part of the TPD of the 2-D BM Eq. (4.22), which is consistent with the fact that in this limit the OU processes Eqs. (4.39) and (4.40) converge to independent GWN processes. It is now worth pointing out that the inhomogeneous process:

$$d\boldsymbol{x} = \sqrt{G(t)}dW_{1,t}, \qquad (4.58)$$

$$dy = \sqrt{G(t)}dW_{2,t}, \qquad (4.59)$$

where $dW_{1,t}$ and $dW_{2,t}$ are two independent GWN processes and

$$G(t) = \sigma^2 + \frac{d}{dt}\left[-\frac{\sigma^2(1 - e^{-\lambda t})}{\lambda}\right] = \sigma^2(1 - e^{-\lambda t}). \qquad (4.60)$$

has a (one-time) probability density similar to Eq. (4.51). Clearly, we have:

$$G(t) < \sigma^2, \quad \forall t \geq 0. \qquad (4.61)$$

We therefore conclude that the more the noise is coloured, the smaller is the effective diffusion coefficient. This result is heuristically expected from the fact that the OU processes are less erratic than the GWN (i.e. note that their realizations of the OU are differentiable) and therefore the exploration process is slower. Remember that the above calculation has however been performed only for the simplest case for which Eqs. (4.37) to (4.40) holds, that is to say when $\delta \equiv 0$ and $R \to \infty$. If a reflecting boundary is imposed on the circle $\partial \mathcal{C}_R$ with $R < \infty$, the qualitative dependence of the effective diffusion on the noise colouration will intuitively remain unchanged. Therefore in view of Eqs. (4.26), (4.35) and (4.61), the insertion mechanism will be less efficient when the band of the excitation noise is reduced.

The above heuristic argumentation suggests that the diminution of the effective diffusion coefficient due to the noise colouration is also expected when non-vanishing headings $\delta(r)$ are considered. It is possible to quantify the effect of noise colouration by using, in the general cases, the approximation schemes developed in [HS], [SA], [HA2]. In view of Eq. (4.27), one would then conclude that $\langle T_a(r_0) \rangle$ strictly decreases for an increase of the noise colouration, (and hence improve the efficiency of the search process). Alternatively, the result given by

Eq. (4.29) is invariable to changes of the colouration of the noise. This can be intuitively understood. Indeed, decreasing the amplitude of the noise while maintaining the strength of the heading, leads to a more rapid approach to the target. By studying the experimental realizations, one has to be aware of the fact that the heading $\delta(r)$ enters into the modeling phenomenologically. This parameter is adjusted to take into account the coupled behaviours of the actuator and the noise generator. The basic philosophy of the random exploration strategy is to select the sizes of the confusion circles $\mathcal{C}_{conf}$ which are:

$$\mathcal{R}_{conf} = \sqrt{\frac{\sigma^2}{2\alpha}} \quad \text{when } \delta = \alpha r \tag{4.62}$$

or

$$\mathcal{R}_{conf} = \frac{\sigma^2}{2\delta} \quad \text{when } \delta = \text{constant.} \tag{4.63}$$

The radius $\mathcal{R}_{conf}$ should be equal to a portion of R, the radius of the exploration disk $\mathcal{C}_R$. Hence, it is $\mathcal{R}_{conf}$ itself that plays the role of the adjustable control parameter of the problem. Imposing that $\mathcal{R}_{conf}$ remain fixed, (i.e. fixing the position of the maximum of $P_{M,S}(r)$), we have: $\alpha = \frac{\sigma^2}{2\mathcal{R}_{conf}^2}$ and respectively $\delta = \frac{\sigma^2}{2\mathcal{R}_{conf}}$. This implies that $\langle T_a(r_0) \rangle$ given by Eqs.(4.27) and (4.29) will exhibit a prefactor proportional to $\frac{\mathcal{R}_{conf}^2}{\sigma^2}$ which is again of the form appearing in Eq. (4.35). Hence, for identical $\mathcal{C}_{conf}$, (i.e. identical location of the maximum of $P_{M,S}(r)$), the increase of the colouration of the noise, (i.e. effectively decreasing σ^2), results in an increase of the mean mating time $\langle T_a(r_0) \rangle$.

4.5 Experimental Results

Figure 4.5 shows an assembly system that comprises a two-degree of freedom fine positioning application, a compliant device and an industrial robot [BD1], [BD2].

The fine positioner is placed on the worktable and carries one of the parts to be mated, while a compliant device, attached to the robot wrist is equipped with a gripper that handles the other part. The system is controlled by a personal computer that communicates with the robot and co-ordinates robot motions with the fine positioner and the compliant device. The fine positioner has been designed to perform a planar and finite band motion with small amplitudes. It is composed of two servo-controlled perpendicular axes actuated by metal bellows cylinders, a technique able to produce the fast and accurate movements necessary to achieve the random search. The operating principle of the system splits the assembly task into two parts. In the first phase, parts are kept in contact while the random planar motion realized by the fine positioner compensates for the matching errors due to system inaccuracies. The second phase is the insertion process itself. The insertion is triggered by the combined actions of the random search mechanism and the maintained contact force between the parts. The noise generator used to activate the fine positioner is a pseudo-random binary sequence

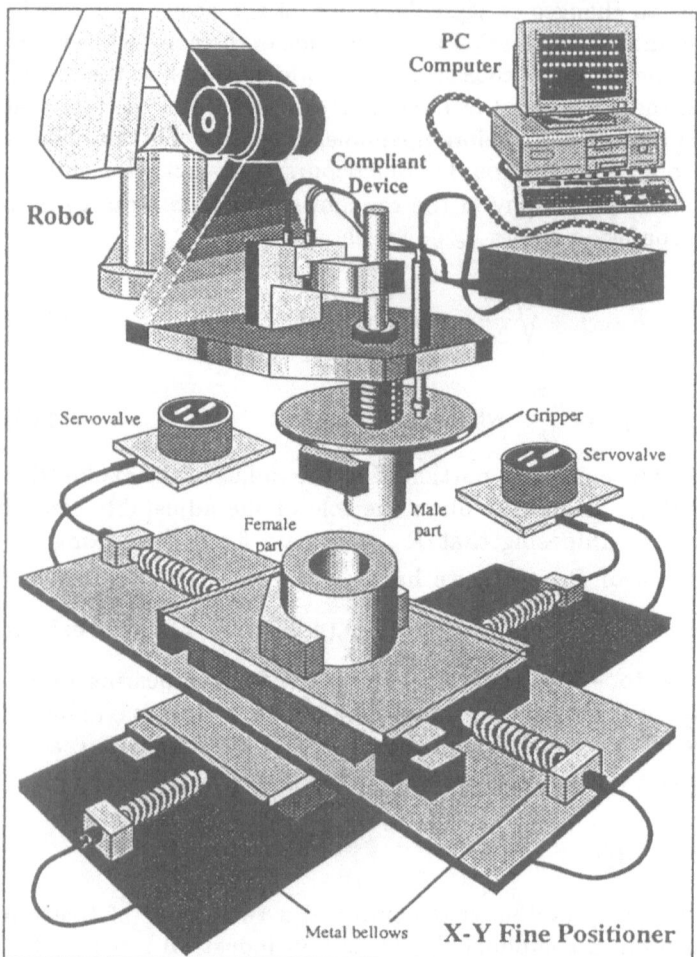

Fig. 4.5. Actual realization of the assembly system.

(PRBS). A PRBS is in fact a two level signal that possesses an autocorrelation function similar to the OU process. By suitably tuning the PRBS generator, it is possible to generate very large band stochastic processes, which, for the typical time scales characterizing the dynamics, are practically GWN. This convergence of the random telegraph process to the GWN is explicitly discussed in [HS]. We have compared the theoretical estimations with actual measurements for the following data [HO3]:

$$a = 0.01 \, [\text{mm}]$$

$$a = 0.03 \, [\text{mm}]$$

$$a = 0.05 \, [\text{mm}]$$

Search times of 900 insertions with initial errors $r_0 = 0.25$, 0.5 and 1.00 [mm] were measured in order to compute the corresponding $\langle T_a(r_0) \rangle$ in the nine cases which result.

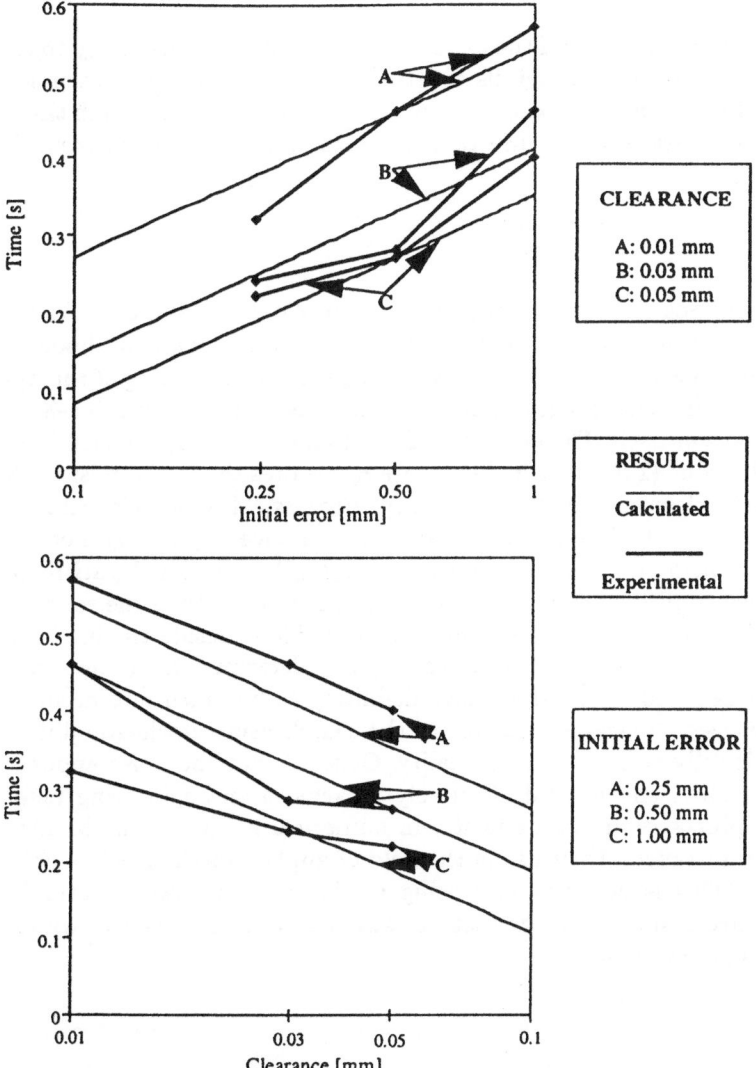

Fig. 4.6. Experimental versus theoretical results for $\langle T_a(r_0) \rangle$, a is the clearance and r_0 the initial error. Note that the scale of the abscissa is logarithmic in the clearance radius parameter, that is, straight lines denote a logarithmic dependency.

To model the dynamics, we have selected the pole-seeking BM with the heading: $\delta = \alpha r$ (i.e. the case c of section 4.2). This choice is based upon the fact that the PRBS which drives the fine positioner is limited both in amplitude and in bandwidth. These characteristics induce the resulting effective movement tendency towards the centre of the searching domain C_R. Moreover, the analysis of the experimental trajectories confirms that their density approximately follows a Gaussian which, for this heading, results in the asymptotic time regime (see Eq. (4.34)).

To compare the theoretical and the experimental results, it is necessary to estimate the values of σ and α. The details of fitting these parameters are discussed in [HO3]. Figure 4.6 presents the experimental and theoretical mean search times that result from an average of 100 insertions for each of the 9 situations considered.

4.6 Conclusions

To solve the circular peg-in-the-hole problem with low clearance and chamferless parts, the use of noisy dynamics as a strategy to explore the neighbourhood of the mating region seems to be particularly well adapted. The modeling of the dynamics by means of diffusion processes enables one to estimate the mean search time needed for the approach. The mean of this random variable $\langle T_a(r_0) \rangle$ exhibits a logarithmic behaviour $\langle T_a(r_0) \rangle \propto \ln(\frac{r_0}{a})$ where r_0 denotes the initial mismatch distance and a is the clearance to allow the mating process. This behaviour has to be contrasted with a purely deterministic search, for instance a spiralling motion, for which a simple estimation leads in this case to $\langle T_a(r_0) \rangle \propto \frac{r_0}{a}$. The logarithmic behaviour has been confirmed experimentally and the superiority of the random search for low clearances has also been practically verified. Beside the intrinsic advantage of the random search, we stress that purely deterministic motions are never strictly realized and fluctuations have definitely to be taken into account for low clearance cases. Hence, a modeling of the search using stochastic differential equations is fully consistent with reality. Observe that the noise appears here as a powerful tool and is used effectively to achieve an engineering task. While stochastic processes are fundamental in numerical methods (the Monte-Carlo methods for instance), their use in the field of applied mechanics has only recently emerged. This is not very surprising as, in the engineering world in general, the presence of noise often constitutes annoyances and nuisances whose influences have to be overcome.

5 Stochastic Buffered Flows

5.1 Introduction

At the level 0 of the general architecture presented in section 1, the flux of matter has to be conveyed between the different work-heads $\{M_k\}$. The most common conveying mechanisms between the machines of a production line can be classified into two main categories: Connections referred to as <u>rigid transfers</u> (RT) and those known as <u>free transfers</u> (FT). In the RT case, if one machine fails, the upstream and downstream fluxes of parts are immediately stopped. This is not the case in FT equipment which allows the fluxes to continue for a while. When the fluxes of matter are very high (i.e. $10^2 - 10^3$ units/ minute), free transfer devices are mandatory. The compressible property in the part-fluxes which characterizes the FT installations is realized by the addition of <u>buffers</u> between the working units. Parts can be stocked in the buffers if it is not possible for them to be processed immediately. Hence, if a machine fails, the upstream one can feed the buffer until it is full (a full buffer implies that the upstream machine is blocked), and the downstream machine can process from the buffer until it is empty (an empty buffer implies that the downstream machine is starved). The study of the buffered flows of matter is an important topic not only in production lines: It also arises in typical areas such as data transmission, traffic problems, dam problems, etc... A recent review on this topic can be found in [DA]. In this chapter we shall perform calculations both for the transient and permanent regimes of the fluxes of matter. The results will be illustrated on an actual line producing cigarettes at a rate of 8000 units/minute. Clearly for such fluxes of matter, a hydrodynamic description is necessary. In this scheme we approximate the population levels (i.e. the number of parts) in the buffers by continuous variables. This is the description we shall adopt here.

5.2 The General Model

The standard scheme of the buffer level dynamics is given by the simple production line schematically represented as:

$$\cdots \longrightarrow M_1 \longrightarrow B_{12} \longrightarrow M_2 \longrightarrow \cdots$$

Two failure prone machines M_1 and M_2 are partly decoupled by the introduction of a buffer B_{12} which has a maximum capacity equals to h [parts]. The
mean time to failure and the mean time to repair will be denoted respectively by
$(\lambda_k)^{-1}$ and $(\mu_k)^{-1}$ for the machine M_k with $k = 1, 2$. The ratio $I_k = \lambda_k(\mu_k)^{-1}$
is the indisposability factor of M_k. The production rate of M_k is C_k [parts/unit
time]. The time dependent population level $X(t)$ of the buffer B_{12} is a continuous
variable, measured from the middle level of B_{12}, i.e. $X(t) \in]-\frac{h}{2}, +\frac{h}{2}[$.

The time evolution of $X(t)$ is described by the stochastic differential equation:

$$\frac{d}{dt}X(t) = C_1\Pi_1(t) - C_2\Pi_2(t) = F(t) \tag{5.1}$$

$$X(t = 0) = x_0 \in]-\frac{h}{2}, +\frac{h}{2}[$$

$$\Pi_k(t) \in \{0,1\}; k = 1, 2, \tag{5.2}$$

where $t \in [0, \infty[$.

In Eqs. (5.1) and (5.2), $\Pi_k(t)$, k=1, 2 are alternating renewal processes [CX]
with the state space $\{0, 1\}$. When M_k is out of order $\Pi_k(t) = 0$ and $\Pi_k(t) = 1$
when M_k is processing parts. When $X(t) = \pm\frac{h}{2}$, some additional information
is needed to characterize the dynamics. In this subsection, we shall restrict our
study to the time intervals for which the boundaries of B_{12} at $\pm\frac{h}{2}$ are not reached.
The behaviour for large times, implying the introduction of the boundary effects
into the dynamics, will be considered later on.

The mean sojourn times in the states $\{0\}$ and $\{1\}$ are respectively $(\mu_k)^{-1}$ and
$(\lambda_k)^{-1}$. The waiting time intervals between transitions from states $\{0\}$ to $\{1\}$
and from $\{1\}$ to $\{0\}$ are characterized respectively by probability distributions
$\psi_k(x)$ and $\phi_k(x)$ on positive definite random variables. Accordingly, we have:

$$\int_0^\infty x\,d\phi_k(x) = (\lambda_k)^{-1}$$

and

$$\int_0^\infty x\,d\psi_k(x) = (\mu_k)^{-1}.$$

An important class of situations is covered by:

$$\phi_k(x) = 1 - \exp(-\lambda_k(x))$$

and

$$\psi_k(x) = 1 - \exp(-\mu_k(x))$$

Indeed, with the use of exponential distributions, $\Pi_k(t)$ are Markovian (di-
chotomous) processes and the noise $F(t)$ in Eq. (5.1) can be represented as a
four state Markov chain with a state space Ω given by:

$$\Omega = \{F_1 = C_1 - C_2, F_2 = -C_2, F_3 = C_1, F_4 = 0\}. \tag{5.3}$$

The transition rates between the F_i's form a 4x4 ergodic matrix $\hat{\mathbf{Q}}$; (in the following, capitals in bold type with a hat denote 4×4 matrices). The matrix $\hat{\mathbf{Q}}$ has the form:

$$\hat{\mathbf{Q}} = \begin{pmatrix} -(\lambda_1 + \lambda_2) & \mu_1 & \mu_2 & 0 \\ \lambda_1 & -(\mu_1 + \lambda_2) & 0 & \mu_2 \\ \lambda_2 & 0 & -(\mu_2 + \lambda_1) & \mu_1 \\ 0 & \lambda_2 & \lambda_1 & -(\mu_1 + \mu_2) \end{pmatrix}. \tag{5.4}$$

Let the 4×4 matrix $\hat{\mathbf{W}}\,(x, t \mid x_0)dx = (W_{ij}(x, t \mid x_0)dx$ be the transition probability density to have $X(t) \in [x, x + dx]$ and $F(t) = F_i$, with the initial conditions $X(t = 0) = x_0$ and $F(t = 0) = F_j$.

It is important here to emphasize that the essential information to characterize the dynamics of $X(t)$ is contained in $\hat{\mathbf{W}}\,(x, t \mid x_0)dx$. We shall calculate this probability density matrix for a restricted time range, of a duration such that $X(t)$ is not allowed to hit one of the boundaries of the buffer B_{12}.

5.3 The Boundary-Free Model and Its Solution

Consider the time t_{dis} defined by:

$$t_{dis} = \frac{min(\frac{h}{2} - x_0, \frac{h}{2} + x_0)}{max(C_1, C_2)} = \frac{min(\frac{h}{2} - x_0, \frac{h}{2} + x_0)}{C_{max}}. \tag{5.5}$$

t_{dis} represents the shortest possible time to completely fill or empty the buffer B_{12}. Obviously, for a fixed x_0, t_{dis} is achieved when the fastest machine is processing parts and the other machine is out of operation. In common applications, the initial condition $X(t = 0) = x_0 = 0$ will be chosen. The stochastic differential Eq. (5.1) with $t \in [0, t_{dis}[$ will be referred to as the boundary-free model. Indeed, in this time range, the boundaries of B_{12} do not affect the dynamics. Let us introduce:

$$P(x, t \mid x_0) = \mathbf{u}^* \hat{\mathbf{W}}(x, t \mid x_0)\mathbf{u} \tag{5.6}$$

and

$$\mathbf{W}(x, t \mid x_0) = \hat{\mathbf{W}}(x, t \mid x_0)\mathbf{u} \tag{5.7}$$

with

$$\mathbf{u}^* = (1, 1, 1, 1) \tag{5.8}$$

where $\mathbf{u}^*$ stands for the transpose of $\mathbf{u}$.

The Chapman-Kolmogorov equation associated with the process $X(t)$ defined by Eq. (5.1) reads [PK], [ZI], [FO], [CL], [MI]:

$$\frac{d}{dt}\mathbf{W}(x, t \mid x_0) = (\hat{\mathbf{F}}\hat{\nabla} + \hat{\mathbf{Q}})\mathbf{W}(x, t \mid x_0) \tag{5.9}$$

with the notation:

$$\hat{\mathbf{F}} = (F_{ij}) = -F_i(\delta_{ij}), \quad i,j = 1,2,3,4. \tag{5.10}$$

and

$$\hat{\nabla} = (\nabla_{ij}) = (\delta_{ij}\frac{\partial}{\partial x}); \quad i,j = 1,2,3,4. \tag{5.11}$$

By direct algebraic elimination, the set of four first order coupled partial differential equations (PDF) Eq. (5.9) can be combined into a single fourth order PDF fulfilled by each component of $\mathbf{W}(x,t \mid x_0)$. This reads:

$$\mathcal{H}(\frac{\partial}{\partial t}, \frac{\partial}{\partial x})W_i(x,t \mid x_0) = 0, \quad i = 1,2,3,4, \tag{5.12}$$

where the differential operator $\mathcal{H}$ is defined by [PK]:

$$\mathcal{H}(\beta,\alpha) = \det(\hat{\mathbf{Q}} - \beta\hat{\mathbf{Id}} + \alpha\hat{\mathbf{F}}) \tag{5.13}$$

where $\hat{\mathbf{Id}}$ is the identity matrix.

From Eqs. (5.6), (5.7) and the linearity of the operator $\mathcal{H}$, we deduce that:

$$\mathcal{H}(\frac{\partial}{\partial t}, \frac{\partial}{\partial x})P(x,t \mid x_0) = 0. \tag{5.14}$$

An explicit expansion of the determinant leads to:

$$\sum_{k=1}^{4} a_k \frac{\partial^k}{\partial t^k} P(x,t \mid x_0) = \sum_{j=0}^{3}\sum_{l=0}^{3} b_{jl} \frac{\partial^j}{\partial t^j}\frac{\partial^l}{\partial x^l} P(x,t \mid x_0), \tag{5.15}$$

with:

$$a_1 = TR,$$
$$a_2 = T^2 + R,$$
$$a_3 = 2T,$$
$$a_4 = 1,$$

where:

$$2T = \text{trace}(\hat{\mathbf{Q}}) \tag{5.16}$$

and

$$R = \lambda_1\lambda_2 + \mu_1\mu_2 + \lambda_1\mu_2 + \lambda_2\mu_1. \tag{5.17}$$

On the right hand side of Eq. (5.15), we calculate explicitly the coefficients b_{jl} that are relevant to the present study, namely:

$$b_{00} = 0 \quad (\Leftarrow \det(\hat{\mathbf{Q}}) \equiv 0) \tag{5.18}$$

$$b_{01} = -TR(\frac{C_1}{1+I_1} - \frac{C_2}{1+I_2}) = -TRv_\infty \tag{5.19}$$

$$I_1 = \frac{\lambda_1}{\mu_1} \text{ and } I_2 = \frac{\lambda_2}{\mu_2}$$

I_1 and I_2 are the indisposability factors and furthermore

$$b_{02} = TR\left(\frac{C_1C_2}{(1+I_1)^2(1+I_2)^2}[\frac{I_1}{\mu_1} + \frac{I_2}{\mu_2} + I_1I_2(\frac{1}{\mu_1} + \frac{1}{\mu_2})]\right) = TR\frac{D}{2} \quad (5.20)$$

The moments $\langle x^m(t) \rangle$ defined by:

$$\langle x^m(t) \rangle = \int_{-\infty}^{+\infty} x^m P(x,t \mid x_0)dx \quad (5.21)$$

can be calculated by multiplying both sides of Eq. (5.15) by x^m and integrating over $]-\infty, +\infty[$. After integrations by parts, we obtain:

$$\sum_{k=1}^{4} a_k \frac{\partial^k}{\partial t^K} \langle x^m(t) \rangle = \sum_{j=0}^{3} \sum_{l=0}^{3} b_{jl} \frac{\partial^j}{\partial t^j}(-1)^l(m)_l \langle x^{m-l}(t) \rangle \quad (5.22)$$

with:

$$(m)_l = \begin{cases} m(m-1)(m-2)...(m-l+1) & \text{if } l \le m, \\ 0, & \text{if } l > m. \end{cases}$$

To solve Eq. (5.22), we need to specify a set of four initial conditions:

$$\frac{\partial^k}{\partial t^k} \langle x^m(t=0) \rangle \quad ; \quad k = 0, 1, 2, 3. \quad (5.23)$$

Using Eq. (5.9), we can write:

$$\frac{\partial^k}{\partial t^k} \int_{-\infty}^{+\infty} x^m P(x,t=0 \mid x_0) = \int_{-\infty}^{+\infty} x^m \mathbf{u}^* \frac{\partial^k}{\partial t^k} \hat{\mathbf{W}}(x,t=0 \mid x_0) \mathbf{u} dx$$

$$= \int_{-\infty}^{+\infty} x^m \mathbf{u}^* (\hat{\mathbf{F}}\hat{\nabla} + \hat{\mathbf{Q}})^k \mathbf{W}(x,t=0 \mid x_0) \quad (5.24)$$

By noting that $\mathbf{u}^*\hat{\mathbf{Q}} = 0$ and introducing the notation:

$$\langle \mathbf{x}^m(0) \rangle = \int_{-\infty}^{+\infty} x^m \mathbf{W}(x,t=0 \mid x_0)dx$$

the integration by parts yields:

$$\langle x^m(0) \rangle = x_0^m, \quad (5.25)$$

$$\frac{\partial}{\partial t}\langle x^m(0) \rangle = -m\mathbf{u}^*\hat{\mathbf{F}}\langle x^{m-1}(0) \rangle = \chi_{1,m}, \quad (5.26)$$

$$\frac{\partial^2}{\partial t^2}\langle x^m(0) \rangle = m(m-1)\mathbf{u}^*\hat{\mathbf{F}}^2\langle x^{m-2}(0) \rangle - m\mathbf{u}^*\mathbf{FQ}\langle x^{m-1} \rangle = \chi_{2,m} \quad (5.27)$$

and

$$\frac{\partial^3}{\partial t^3}\langle x^m(0)\rangle = -m(m-1)(m-2)\mathbf{u}^*\mathbf{\hat{F}}^3\langle x^{m-3}(0)\rangle$$

$$+ m(m-1)\mathbf{u}^*\mathbf{\hat{F}}(\mathbf{\hat{F}\hat{Q}} + \mathbf{\hat{Q}\hat{F}})\langle x^{m-2}(0)\rangle - m\mathbf{u}^*\mathbf{\hat{F}}\mathbf{\hat{Q}}^2\langle x^{m-1}(0)\rangle = \chi_{3,m}. \quad (5.28)$$

Eq. (5.22) and Eqs. (5.25) to (5.28) fully characterize the dynamics of the moments of the process $X(t)$ for $t < t_{dis}$.

The characteristic equation associated with the left hand side of Eq. (5.22) is a cubic equation:

$$\sum_{k=1}^{4} a_k \omega^{k-1} = 0 \quad (5.29)$$

and its discriminant is given by:

$$\Delta = -(108)R^2(\lambda_1 - \lambda_2 + \mu_1 - \mu_2)^2. \quad (5.30)$$

Therefore, the solutions of Eq. (5.29) are real and from Lemma 2 of [PK], they are also negative. We shall write:

$$-\omega_3 \leq -\omega_2 \leq -\omega_1 < 0. \quad (5.31)$$

These roots distinguish the two regimes in the evolution of the moments, namely:

1) The short transient regime.
This is achieved when $t \ll (\omega_3)^{-1}$. Here Eqs. (5.25) to (5.28) lead to:

$$\langle x^m(t)\rangle \simeq x_0^m + \chi_{1,m}t + \chi_{2,m}\frac{t^2}{2} + \chi_{3,m}\frac{t^3}{6}. \quad (5.32)$$

2) The asymptotic regime.
This regime is reached when $t \gg (\omega_1)^{-1} = t_{relax}$, provided the buffer capacity h is large enough to allow $t_{relax} << t_{dis}$. Here, the behaviour is determined by the special solution of the ordinary differential Eq. (5.22). We obtain:

$$\langle x(t)\rangle \simeq \frac{-b_{01}}{a_1} = v_\infty t \quad (5.33)$$

and

$$\langle x^2(t)\rangle \simeq 2\frac{b_{02}}{a_1} = Dt \quad (5.34)$$

where v_∞ and D are defined in Eqs. (5.19) and (5.20) respectively.

It is important to remark that in the time asymptotic regime, if the capacity of the buffer is such that Eq. (5.5) is satisfied, the process $X(t)$ exhibits a diffusive behaviour. This is a consequence of the central limit theorem which, in this context, is discussed in [PK].

5.4 Examples

To further explore the dynamics, let us be more specific about the initial condition $\langle \mathbf{x}^m(0) \rangle$. We will assume that at $t = 0$, both machines M_1 and M_2 are in operation, (i.e. $F(t = 0) = F_1 = C_1 - C_2$) and that the buffer is half populated, i.e. $x(0) = 0$. In terms of probability densities, we have:

$$\mathbf{W}(x, t = 0 \mid x_0) = \delta(x)(1, 0, 0, 0)^* \tag{5.35}$$

and therefore

$$\langle \mathbf{x}^m(0) \rangle = \begin{cases} (1, 0, 0, 0)^* & \text{if } m = 0; \\ (0, 0, 0, 0)^* & \text{if } m > 0. \end{cases} \tag{5.36}$$

Eqs. (5.25) to (5.28) together imply that:

$$\langle x^m(0) \rangle = 0, \quad m > 0. \tag{5.37}$$

$$\frac{\partial}{\partial t} \langle x^m(0) \rangle = \begin{cases} (C_1 - C_2), & m{=}1; \\ 0, & m \geq 2. \end{cases} \tag{5.38}$$

$$\frac{\partial^2}{\partial t^2} \langle x^m(0) \rangle = \begin{cases} C_2 \lambda_2 - C_1 \lambda_1, & m{=}1; \\ 2(C_1 - C_2)^2, & m{=}2; \\ 0, & m > 2. \end{cases} \tag{5.39}$$

$$\frac{\partial^3}{\partial t^3} \langle x^m(0) \rangle = \begin{cases} \Gamma_1, & m{=}1; \\ \Gamma_2, & m{=}2; \\ -6(C_1 - C_2)^3, & m{=}3; \\ 0, & m > 3. \end{cases} \tag{5.40}$$

with:

$$\Gamma_1 = (C_1 - C_2)[(\lambda_1 - \lambda_2)^2 + \lambda_1 \mu_1 + \lambda_2 \mu_2] +$$

$$+ \mu_1 \lambda_1 C_2(1 + I_1) - \mu_2 \lambda_2 C_1(1 + I_2), \tag{5.41}$$

and

$$\frac{1}{2}\Gamma_2 = \lambda_1 C_2^2 + \lambda_2 C_1^2 -$$

$$- (C_1 - C_2)[2(C_1 - C_2)(\lambda_1 + \lambda_2) - \lambda_1 C_2 + \lambda_2 C_1]. \tag{5.42}$$

Let us now present the results obtained for two particular configurations.

a) Different machines with identical indisposabilities and processing rates.

$$C_1 = C_2 = C,$$

$$\lambda_2 = 2\lambda_1$$
$$\mu_2 = 2\mu_1$$

and therefore
$$I_1 = I_2$$

In this case, we obtain:
$$\omega_3 = 3\gamma_1$$
$$\omega_2 = 2\gamma_1$$
$$\omega_1 = \gamma_1$$
$$\gamma_1 = (\lambda_1 + \mu_1).$$

The use of Eq. (5.22) together with Eqs. (5.37) to (5.40) yields:

$$\langle x(t) \rangle = \frac{C\lambda_1}{\gamma_1^2}\left(-e^{-\gamma_1 t} + \frac{1}{2}e^{-2\gamma_1 t} + \frac{1}{2}\right), \qquad (5.43)$$

and

$$\langle x^2(t) \rangle = \frac{C^2\lambda_1}{2\gamma_1^4}\Big((12\mu_1 - 6\lambda_1)e^{-\gamma_1 t} + (6\lambda_1 - 3\mu_1)e^{-2\gamma_1 t} - 2\lambda_1 e^{-3\gamma_1 t} +$$

$$+ 6\mu_1\gamma_1 t - 9\mu_1 + 2\lambda_1\Big). \qquad (5.44)$$

b) Identical machines.

$$C_1 = C_2 = C$$
$$\lambda_1 = \lambda_2 = \lambda$$
$$\mu_1 = \mu_2 = \mu.$$

In this case, the discriminant given by Eq. (5.30) is $\Delta = 0$. Hence we can deduce that there are two roots of the cubic equation that are equal. We obtain:

$$\omega_3 = 2\gamma$$

$$\omega_2 = \omega_1 = \gamma$$

$$\gamma = (\lambda + \mu) = \mu(1 + I).$$

This leads to:

$$\langle x(t) \rangle = 0 \tag{5.45}$$

and

$$\langle x^2(t) \rangle = D\left[\left(\frac{2}{\gamma} + (1+I)t\right)e^{-\gamma t} - \frac{I}{2\gamma}(e^{-2\gamma t} - 1) + t - \frac{2}{\gamma} \right], \tag{5.46}$$

where, according to Eq. (5.20):

$$D = \frac{4C^2 \lambda \mu}{\gamma^3}. \tag{5.47}$$

For identical machines, we remark that Eq. (5.15) takes the simple form:

$$\Theta(\Theta - \gamma^2)P(x,t \mid x_0) = [\Theta - (\lambda - \mu)^2]C^2 \frac{\partial^2}{\partial x^2} P(x,t \mid x_0), \tag{5.48}$$

with

$$\Theta = (\frac{\partial}{\partial t} + \gamma)^2. \tag{5.49}$$

Eq. (5.48) is quadratic in the operator Θ. This enables us to calculate explicitly the Fourier transform of $P(x,t \mid x_0)$:

$$P(k,t \mid x_0) = \int_{-\infty}^{+\infty} e^{-ikx} P(x,t \mid x_0)dx$$

$$= e^{-\gamma t}\Big[A_1(k)\cosh(\Omega_1(k)t) + A_2(k)\cosh(\Omega_2(k)t) +$$

$$+ B_1(k)\sinh(\Omega_1(k)t) + B_2(k)\sinh(\Omega_2(k)t) \Big] \tag{5.50}$$

where

$$\Omega_1 = \sqrt{f + \sqrt{f^2 + g^2}}, \tag{5.51}$$

$$\Omega_2 = \sqrt{f - \sqrt{f^2 + g^2}}, \tag{5.52}$$

and

$$f = \frac{1}{2}(\gamma^2 - C^2 k^2), \tag{5.53}$$

$$g = C^2 k^2 (\lambda - \mu)^2. \tag{5.54}$$

The amplitudes $A_1(k), A_2(k), B_1(k)$ and $B_2(k)$ can be determined from the initial conditions Eqs. (5.25) to (5.28).

5.5 Discussion of the Transient Analysis of the Boundary-Free Model

The results obtained in sections 5.3 and 5.4 are exact as long as $t < t_{dis}$. If $t > t_{dis}$, it is then possible that $X(t)$ can hit one of the boundaries of B_{12}. Once a boundary of B_{12} has been touched, the dynamical behaviour is modified according to the operating conditions for the system with empty or full buffers. In the approach adopted here, the time evolution of the moments of the population level in a buffer separating two failure-prone machines involves two essential hypotheses.

a) The operating states of the machines are governed by Markovian alternating renewal processes; (i.e. the distributions $\phi_k(x)$ and $\psi_k(x)$ are exponentials).

b) The buffer level is represented by a continuous variable.

Hypothesis a) may seem at first sight as rather restrictive, as pure Markovian alternating processes can barely be expected in actual production lines. As an immediate consequence of the Markov assumption, we no longer have the freedom of introducing arbitrary coefficients of variation (CV) for the times to failure and the times for repairs. Remember that the CV is the ratio between the centred second moment and the square of the mean. It is easy to verify that the exponential distributions yield CV identically equal to unity. When the probabilistic behaviour of the machines is not Markovian (this is always the case when the CV is not unity), and large buffer capacities are available, a simple argument enables one to predict qualitatively the modified time evolution. Indeed, with large buffer capacities, the population level process always relaxes to a diffusive regime for any of underlying alternating renewal processes describing the operating states of the machines. This is a direct consequence of the Central Limit Theorem. The relaxation time and the final diffusion constant are obviously dependent on the distributions $\phi_k(x)$ and $\psi_k(x)$, and hence on their coefficients of variation. In particular, the larger the coefficients of variation are the larger are the diffusion constants. Explicit calculations using phase type distributions [NE] can be performed. For the situation where one machine is ideal ($I = 0$) and the other is governed by distributions $\psi_k(x)$ and $\phi_k(x)$ of the Erlang type, an explicit expression is obtained in [HO2]. Let us mention that more general distributions (i.e. phase-type) are also introduced in [JF] in the discrete modeling framework, (i.e. the population of the buffer is a discrete variable).

Hypothesis b) is in fact rather mild. Indeed when the buffer capacity is large and the flux of matter is high, the hydrodynamical model is fully justified [DA]. Moreover, as a recent study [DV] shows, the hydrodynamic model can be used to approach the behaviour of a discrete level model by a suitable buffer capacity renormalization. Note that a large number of authors have made contributions devoted to the discussion of the dynamics of the buffer population with discrete variables. We refrain here from exhaustively listing these researches. Let us nevertheless quote the seminal papers by [GE], [BU] and refer to [DA] for a complete bibliography. Note that from an analytical viewpoint, the description in terms of discrete variables for the population level is less attractive. Indeed, one rapidly has to handle Markov chains with a large number of states. In contrast, in the hydrodynamic framework one reduces the discussion to a linear field theory.

Let us now assume that $x_0 = 0$. Hence we start wih a half filled buffer. The production line is assumed to be balanced i.e. $v_\infty = 0 \iff \frac{C_1}{1+I_1} = \frac{C_2}{1+I_2}$.

a) Behaviour of the mean.

From Eq. (5.33), we observe that the $\lim_{t\to\infty} X(t) = x_\infty$ exists. This can be intuitively expected from the balancing of the production of M_1 and M_2. Observe however, that before reaching this time independent regime, $\langle X(t) \rangle$ moves from its initial position $X(0) = 0$. From the explicit solution given in Eqs. (5.43) and (5.45) we indeed have:

$$x_\infty = \frac{C\lambda_1}{2\gamma_1^2} \neq x_0 = 0 \qquad (5.55)$$

which holds for different machines with identical indisposabilities and processing rates. For perfectly identical machines, we have:

$$x_\infty = x_0 = 0 \qquad (5.56)$$

as is intuitively expected from symmetry arguments.

Therefore, to prevent a systematic mean approach towards a full or empty buffer, the capacity h has to be selected such that $x_\infty \in \,] - \frac{h}{2}, +\frac{h}{2} [$.

b) Behaviour of $\langle x^2(t) \rangle$ and the first access time to the boundaries of B_{12}.

Of course, the mean first access time $\langle T \rangle$ to one of the boundaries $\pm\frac{h}{2}$ of B_{12} is an important quantity to calculate. An estimation of $\langle T \rangle$ can be obtained as follows:

$$\langle x^2(\langle T \rangle) \rangle = \frac{h^2}{4}, \qquad (5.57)$$

which simply expresses the fact that at $\langle T \rangle$, the amplitude of the fluctuations equals the size of the buffer capacity. For instance, using Eqs. (5.44) or (5.46), $\langle T \rangle$ obeys transcendental equations only solvable by numerical means. For large buffer capacities, implying $t_{dis} >> t_{relax} = (\omega_1)^{-1}$, $\langle T \rangle$ can be estimated by the following expression:

$$\langle T \rangle \approx (t_{mtf} + t_{fp}), \qquad (5.58)$$

where

$$t_{mtf} = (\lambda_1 + \lambda_2)^{-1} \qquad (5.59)$$

is the mean time to failure of one of the machines M_1 or M_2 and

$$t_{fp} = \frac{h^2}{4D}. \qquad (5.60)$$

Eq. (5.60) provides the mean first passage time to $\pm\frac{h}{2}$ for a pure Brownian motion with variance D, [KA]. Remember that to write Eq. (5.58), we have implicitly assumed that at time $t = 0$, both machines are operating.

Let us remark that an exact expression for $\langle T \rangle$ is available for the simplest situation encountered when one of the machines is ideal, i.e. when one of the indisposabilities I_k vanishes. In this case, we can directly use the results obtained for the mean first access time to $\pm\frac{h}{2}$ given in [MS], [HA1].

For instance assume that $I_1 = 0$ and that the line is balanced. In other words we assume: $C_1 = \frac{C_2}{1+I_2}$. Starting with $X(t = 0) = 0$ and with M_2 operating, the formula Eq. (17) of [MS] provides an exact expression for $\langle T \rangle$ in the form:

$$\langle T \rangle = \frac{h^2}{4D} + \frac{F}{2\lambda_2}\left(\frac{1 + \frac{F}{2}(1 + I_2)}{1 + F}\right)$$

$$= \frac{F^2}{8\lambda_2}(1 + I_2) + \frac{F}{4\lambda_2}\left(\frac{1 + \frac{F}{2}(1 + I_2)}{1 + 2F}\right), \tag{5.61}$$

where the dimensionless parameter F is defined by:

$$F = \frac{\mu_2 h}{C_1} \tag{5.62}$$

and the diffusion coefficient is given here by:

$$D = \frac{C_1^2 \lambda_2}{2\mu_2^2(1 + I_2)}. \tag{5.63}$$

Observe that for a large buffer capacity and therefore for a large F, the diffusion contribution t_{fp} in Eq. (5.61) is dominant in the expression for $\langle T \rangle$.

5.6 Case Study from a Cigarette Production Unit

Let us apply the above results to a concrete situation encountered in a cigarette production unit. The architecture of the unit is composed of two machines as sketched in section 5.1. The machine M_1 is processing the cigarettes and the machine M_2 is packing the output of M_1. The dipole M_1 and M_2 includes a buffer B_{12}, (B_{12} is known as the "central lung" in the factory!). For this situation, the following data was available.

$$h = 6.5 \times 10^4 \text{ [cigarettes]},$$

$$C_1 = C_2 = C = 8 \times 10^3 \text{ [cigarettes/minute]},$$

$$\lambda_1 = 10^{-3} \text{ [sec]}^{-1},$$

$$\mu_1 = 5 \times 10^{-3} \text{ [sec]}^{-1},$$

$$\lambda_2 = 2\lambda_1,$$

$$\mu_2 = 2\mu_1.$$

. From these data we have $I_1 = I_2 = I = 0.2$. It is assumed that $x_0 = 0$ at time t=0. This data explicitly shows that the line is balanced in the mean, (i.e. $C_1(\frac{1}{1+I_1}) = C_2(\frac{1}{1+I_2})$. The distributions $\phi_k(x)$ and $\psi_k(x)$ were unknown. In particular, the coefficients of variation were unfortunately not recorded. The very high production rate and the large buffer capacity fully justify the use of the hydrodynamical representation. A large buffer size was easy to implement in this case as the price and the volume of the buffer space are, in this example, pretty low. Assuming that the failure and the repair times are governed by Markov alternating processes, the results of section 5.5 imply:

$$t_{dis} = \frac{3.25 \times 10^4}{8 \times 10^3} \approx 4\,[\text{minutes}],$$

$$t_{relax} = \frac{1}{(\lambda_1 + \mu_1)} \approx 2\frac{3}{4}\,[\text{minutes}],$$

$$t_{mtf} = (\lambda_1 + \lambda_2)^{-1} \approx 5\frac{1}{2}\,[\text{minutes}],$$

$$D = \frac{3C^2\lambda}{(1+I)^3\mu^2} \approx 1.25 \times 10^6[\text{cigarettes}^2/\text{sec}],$$

such that:

$$t_{fp} = \frac{h^2}{4D} = \frac{(6.5)^2 \times 10^8}{4 \times 1.25 \times 10^6} \approx 11\frac{2}{3}[\text{minutes}].$$

$$x_\infty = \frac{C_1\lambda_1}{2\gamma_1^2} = \frac{8 \times 10^3 \times 10^{-3}}{2 \times 60 \times (6 \times 10^{-3})^2} \approx 1.85 \times 10^3\,[\text{cigarettes}].$$

Observe that the buffer capacity is chosen here to yield $t_{dis} > t_{relax}$ and hence $X(t)$ essentially reaches its diffusive regime before the boundaries $\pm\frac{h}{2}$ are hit. Note also that $x_\infty << h$, therefore the mean tendency to drift to one of the boundaries of B_{12} is avoided for a large portion of the possible starting conditions x_0, (i.e. $x_0 \in [-\frac{h}{2},(\frac{h}{2} - x_\infty)]$. In view of Eq. (5.58), we have:

$$\langle T \rangle \approx (t_{mtf} + t_{fp}) \approx 17\,[\text{minutes}]$$

a value which corresponds to the operators' observations, although $\langle T \rangle$ has not yet been systematically recorded.

5.7 Permanent Regimes

So far we have restricted our study to times shorter than t_{dis} defined by Eq. (5.5). In this section, we shall discuss the expected throughput in the stationary regime (i.e. when $t \to \infty$). To this end, we obviously have to take into account the effects of the boundaries of the buffer. Hence, Eq. (5.9) has to be supplemented with a description of boundary conditions. For simplicity of treatment, we restrict our attention, from now on, to two identical machines (C, λ, μ). Refer to [CL] for the more general cases. We can write:

$$\frac{d}{dt}\mathbf{W}(x,t\mid x_0) = (\hat{\mathbf{F}}\hat{\nabla} + \hat{\mathbf{Q}})\mathbf{W}(x,t\mid x_0) + \mathbf{B}(t) \qquad (5.64)$$

where Eq. (5.64) is valid only when $\frac{-h}{2} < X(t) < \frac{h}{2}$ and the vector $\mathbf{B}(t)$ has the form:

$$\mathbf{B}(t)^* = (0, CW_2(x = -\frac{h}{2}, t), CW_3(x = +\frac{h}{2}, t), 0). \qquad (5.65)$$

Let $m_{i,j}$ be the Prob$\{X(t) = -\frac{h}{2}$ and $\Pi_1(t) = i$ and $\Pi_2(t) = j\}$ and $M_{i,j}$ be the Prob$\{X(t) = \frac{+h}{2}$ and $\Pi_1(t) = i$ and $\Pi_2(t) = j\}$. Note that $m_{1,0}(t) \equiv 0$ and $M_{0,1}(t) \equiv 0$ as they both correspond to unstable states. We denote:

$$\mathbf{m}^*(t) = (m_{0,0}(t), m_{0,1}(t), m_{1,0}(t), m_{1,1}(t)) \qquad (5.66)$$

and similarly

$$\mathbf{M}^*(t) = (M_{0,0}(t), M_{0,1}(t), M_{1,0}(t), M_{1,1}(t)). \qquad (5.67)$$

Now we have to distinguish between the two different boundary behaviours which result from the possibility that a starved or blocked machine has to fail or not. These boundary behaviours will be referred to as OD for the operation dependent failures (i.e. those failures which occur *only when a machine is operating*) and by TD for the time dependent failures which occur *even if a machine is starved or blocked* due to an empty upstream buffer or a full downstream buffer. The population dynamics of $\mathbf{m}(t)$ and $\mathbf{M}(t)$ are given by the rate equations (the subscripts OD or TD stand for the failure mechanisms under study):

$$\frac{\partial}{\partial t}\mathbf{m}(t) = \hat{\Lambda}_{1,\kappa}\mathbf{m} + \mathbf{B}_1(t) \quad \kappa = \text{OD or TD} \qquad (5.68)$$

and

$$\frac{\partial}{\partial t}\mathbf{M}(t) = \hat{\Lambda}_{2,\kappa}\mathbf{M} + \mathbf{B}_2(t) \quad \kappa = \text{OD or TD}, \qquad (5.69)$$

with:

$$\hat{\Lambda}_{1,OD} = \begin{pmatrix} -2\mu & 0 & 0 & 0 \\ \mu & -\mu & 0 & \lambda \\ 0 & 0 & 0 & 0 \\ 0 & \mu & 0 & -2\lambda \end{pmatrix} \qquad (5.70)$$

or

$$\hat{\Lambda}_{1,TD} = \begin{pmatrix} -2\mu & \lambda & 0 & 0 \\ \mu & -(\mu+\lambda) & 0 & \lambda \\ 0 & 0 & 0 & 0 \\ 0 & \mu & 0 & -2\lambda \end{pmatrix} \qquad (5.71)$$

and

$$\hat{\Lambda}_{2,OD} = \begin{pmatrix} -2\mu & 0 & 0 & 0 \\ 0 & 0 & 0 & 0 \\ \mu & 0 & -\mu & \lambda \\ 0 & \mu & 0 & -2\lambda \end{pmatrix} \qquad (5.72)$$

or

$$\hat{\Lambda}_{2,TD} = \begin{pmatrix} -2\mu & 0 & \lambda & 0 \\ 0 & 0 & 0 & 0 \\ \mu & 0 & -(\mu+\lambda) & \lambda \\ 0 & \mu & 0 & -2\lambda \end{pmatrix} \tag{5.73}$$

and the boundary terms $\mathbf{B}_1(t)$ and $\mathbf{B}_2(t)$ read:

$$\mathbf{B}_1(t)^* = (0, CW_2(x = -\frac{h}{2}, t), 0, 0) \tag{5.74}$$

and

$$\mathbf{B}_2(t)^* = (0, 0, CW_3(x = +\frac{h}{2}, t), 0). \tag{5.75}$$

As this section deals with the stationary regimes, we therefore solve the time-independent Eqs. (5.68) and (5.69) in both the OD and TD situations.

a) Operation dependent failing machines. In this case, we obtain:

$$W_4 = \frac{\lambda}{\mu}K \;\; ; \;\; W_3 = W_2 = K \;\; ; \;\; W_1 = \frac{\mu}{\lambda}K, \tag{5.76}$$

$$m_{0,0} = m_{1,0} = 0 \;\; ; \;\; m_{0,1} = K\frac{2C}{\mu} \;\; ; \;\; , m_{1,1} = K\frac{C}{\lambda} \tag{5.77}$$

and

$$M_{0,0} = M_{0,1} = 0 \;\; ; \;\; M_{1,0} = K\frac{2C}{\mu} \;\; ; \;\; M_{1,1} = K\frac{C}{\lambda}, \tag{5.78}$$

where K is the normalization constant which in this case reads:

$$K^{-1} = h(\frac{(\lambda+\mu)^2}{\lambda\mu}) + 2C\frac{2\lambda+\mu}{\lambda\mu}. \tag{5.79}$$

The throughput Z_{OD}, (i.e. the output of the downstream machine M_2) is given by the expression:

$$Z_{OD} = C(W_1h + W_2h + m_{1,1} + M_{1,1}). \tag{5.80}$$

In view of Eqs. (5.80) and (5.76)-(5.79), we end with:

$$Z_{OD} = C(\frac{1}{1 + I_{\text{eff}}}) \tag{5.81}$$

with

$$I_{\text{eff}} = I(1 + \frac{1}{1 + \frac{F}{2}(1 + I)}) \;\; ; \;\; F = \frac{\mu h}{C}. \tag{5.82}$$

b) Time dependent failure machines. Here, we reobtain Eq. (5.76) which does not depend on the boundary behaviour and

$$m_{0,0} = K\frac{\lambda C}{\mu(\lambda+\mu)} \;\; ; \;\; m_{0,1} = K\frac{2C}{\lambda+\mu} \;\; ; \;\; m_{1,0} = 0 \;\; ;$$

$$m_{1,1} = K \frac{\mu C}{\lambda(\lambda + \mu)} \tag{5.83}$$

and

$$M_{0,0} = K \frac{\lambda C}{\mu(\lambda + \mu)} \; ; \; M_{0,1} = 0 \; ; \; M_{1,0} = K \frac{2C}{\lambda + \mu} \; ,$$

$$M_{1,1} = K \frac{\mu C}{\lambda(\lambda + \mu)}. \tag{5.84}$$

Accordingly, the normalization factor K in this case reads:

$$K = \frac{(\lambda + \mu)^2}{\lambda \mu} [h + \frac{2C}{\lambda + \mu}]. \tag{5.85}$$

Finally, the throughput, Z_{TD}, in this case reads:

$$Z_{\text{TD}} = C (W_1 h + W_2 h + m_{1,1} + M_{1,1}) =$$

$$= C (\frac{1}{1+I})^2 (1 + I(\frac{F}{F + \frac{2}{1+I}})) \tag{5.86}$$

with the indisposability factor $I = \frac{\lambda}{\mu}$ and F as defined in Eq. (5.82). The result given in Eq. (5.86) can be found in [WG].

Note that when $F \to \infty$ the throughput is in both cases given by Eqs. (5.82) and (5.86) of the form $Z_{\text{TD}} = Z_{\text{OD}} = C(\frac{1}{1+I})$. This result can be understood if one notes that generally in the most favourable case, (i.e. when $F \to \infty$), the mean throughput equals at most to the mean throughput of the machine presenting the highest performance.

The behaviour of Z_{TD} and Z_{OD} is qualitatively described in Figure 5.1.

Clearly, $Z_{\text{TD}} \leq Z_{\text{OD}}$ as expected from intuition (i.e. for identical indisposability factors, TD failures are more frequent than OD failures). Remark that for $F \to 0 \Rightarrow h \to 0$ $Z_{\text{TD}} = C(\frac{1}{1+I})^2$ which simply expresses that the reliability of the dipole is the product of the reliabilities of two independent machines as is the case for the TD failing mechanisms with no buffer.

For cases in which the machines M_1 and M_2 are different, more complicated expressions result, some of these cases are discussed in [CL], [DB], [WG].

5.8 Aggregation Method

The permanent regime analysis has only been performed for a dipole configuration, (i.e. two machines separated by one buffer). In general however, the production line is composed of several dipoles. For a long chain, the exact results are very complex to develop. Hence, one relies on approximation techniques which mainly consist of using the results obtained for the dipole configuration. There is an vast amount of work devoted to this approximation scheme and the reader is invited to consult [DA] for a complete review. A detailed analysis valid for the hydrodynamical representation is given by [TE]. Here we shall confine our analysis to the rather ideal case where all machines are identical and for machines failing only when operating, (i.e. OD failure-prone machines). This ideal

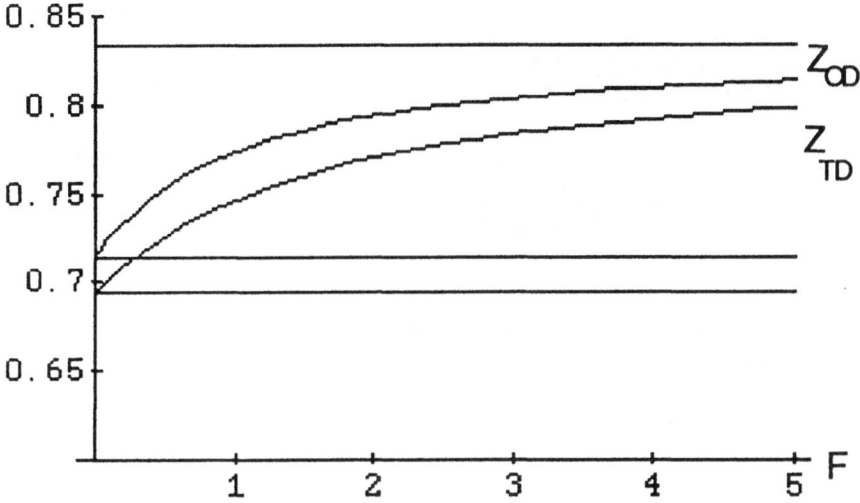

Fig. 5.1. Comparison of the throughput for the time dependent and the operation dependent failures.

configuration can however always be used to obtain the lower and upper bounds of a more general situation. Indeed, assuming that all machines are identical to the worst operating one will definitely produce a lower estimate of the general throughput and conversely if the line is approximated by taking machines identical to the best performing one, we will then obtain an optimistic bound.

The basis of the method to handle long production lines is sketched in Figure 5.2 where the aggregation procedure is summarized. The procedure is to consider a dipole configuration as behaving as a single effective machine and then to reduce iteratively the number of stations until only one effective machine is obtained at the end of the procedure.

This technique is only approximate as it lumps a four state Markov chain, (two machines are coupled), into a two state one which describes the resulting effective machine. Moreover, when machines are not identical, the effective machine obtained after the aggregation procedure can behave differently if it is regarded from upstream or downstream. Hence, the aggregation procedure depends in general on the order in which the compactification is performed (see [TE] for a detailed discussion). When the machines are identical, the situation is simpler and the results obtained in [TE] for this particular case, can be directly read in Eq. (5.81) which is the throughput for a dipole with machines failing according to the operation dependent principle. For identical machines, the order of the aggregation is irrelevant [TE]. In this case, we observe that the throughput of the dipole given by Eq. (5.81) can be written in terms of an effective indisposability factor I_{eff}. Hence, Eq. (5.81) can be directly interpreted as the result obtained after one stage of the aggregation procedure. Remark similarly, that if TD failures are considered, one can use Eq. (5.86) as the starting point of an

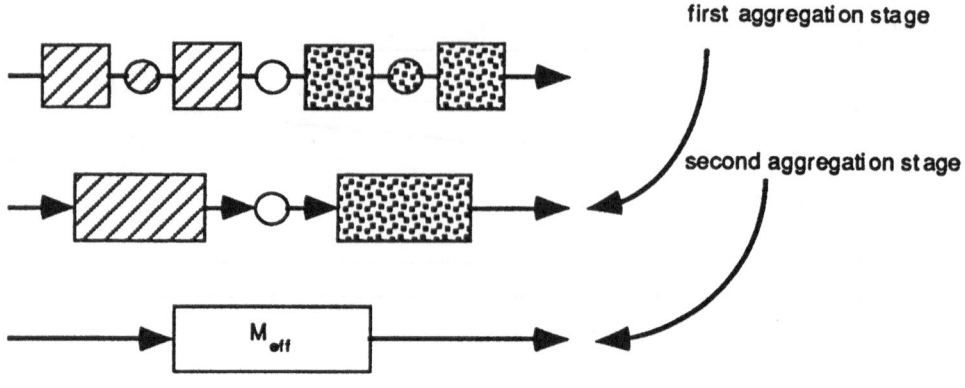

Fig. 5.2. Sketch of the basic aggregation method. Each dipole composed of two machines separated by one buffer is assimilated to one effective machine.

aggregation procedure valid for identical machines. For non-identical machines, the elegant form given by Eq. (5.81) is no longer valid [TE], [CL].

To illustrate the use of the method let us consider a special problem which has been encountered in real situations [FI].

Problem.

Consider a production line composed of four identical machines M_k, $k = 1, 2, ...$ placed in series. The time between failure and the repair times are exponentially distributed with parameters λ and μ respectively. To improve the throughput of the line, we equip the system with buffers with a total capacity of H parts. Two configurations of the line are to be studied and are sketched in Figures 5.2 and 5.3. Which configuration yields the best throughput?

a) Throughput Z_a with central buffer, (Figure 5.3).

The aggregation is conducted in two stages as represented in Figure 5.2.

After the first step machines M_1 and M_2 are aggregated into machines M_{12} and similarly with machines M_3 and M_4 into the resulting effective machine M_{34}. Machines M_{12} and M_{34} have an effective indisposability $2I = 2\frac{\lambda}{\mu}$. The final aggregation to the effective machine M_{1234} is conducted between the two identical machines M_{12} and M_{34} and therefore using Eqs. (5.81) and (5.82), we are led to:

$$Z_a = C\frac{1}{1 + I_{\text{eff}}} \quad I_{\text{eff}} = 2I\left(1 + \frac{1}{1 + \frac{3}{2}(1 + 2I)F}\right) = C\frac{1}{1 + I_a}, \qquad (5.87)$$

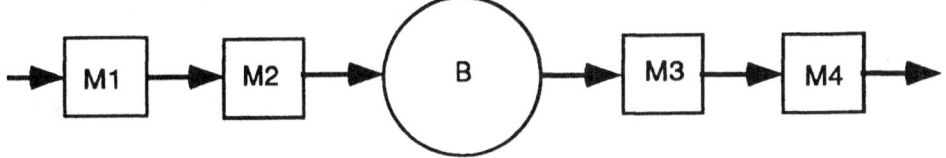

Fig. 5.3. Configuration of a four machine production line with a central buffer, (central lung solution). The throughput resulting from this configuration is denoted Z_a

with the dimensionless parameter F defined by:

$$F = \frac{\mu H}{3C}.\qquad(5.88)$$

b) Throughput Z_b with equidistributed buffers, (Figure 5. 4).

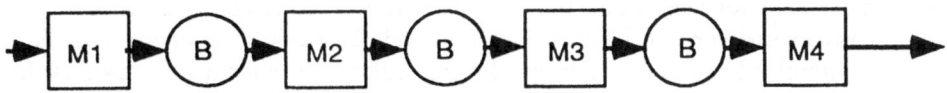

Fig. 5.4. Configuration of a four machine production line with equidistributed buffer places. The throughput resulting from this configuration is denoted Z_b

When the buffer places are distributed among three equal buffers, the aggregation is again performed in two stages and using Eq. (5.81), we obtain:

$$Z_b = C\frac{1}{1 + I_{\text{eff},1}\left(1 + \frac{1}{\frac{F}{2}(1 + I_{\text{eff},1})}\right)} = C\frac{1}{1 + I_b},\qquad(5.89)$$

with:

$$I_{\text{eff},1} = I\left(1 + \frac{1}{1 + \frac{1}{2}(1 + I)F}\right) \approx 2I(1 - \frac{F}{4}(1 + I))\qquad(5.90)$$

and F is defined as before in Eq. (5.88).

For small buffer capacities, (i.e. $F \ll 1$), the throughput Z_a and Z_b are expanded into first order in F and we obtain:

$$Z_a = \frac{1}{1 + I_a}, \quad I_a = 4I(1 - \frac{3F}{4}(1 + 2I) + o(f) \tag{5.91}$$

and

$$Z_b = \frac{1}{1 + I_b}, \quad I_b = 4I(1 - \frac{F}{2} - \frac{3}{4}FI) + o(f). \tag{5.92}$$

Hence, we deduce immediately :

$$Z_a > Z_b, \quad \text{for } F \ll 1. \tag{5.93}$$

On the other extreme, when $F \to \infty$ we have:

$$Z_a \to C\frac{1}{1 + 2I}, \quad \left(I = \frac{\lambda}{\mu}\right) \tag{5.94}$$

and

$$Z_b \to C\frac{1}{1 + I} \tag{5.95}$$

Hence when F is large, the configuration with equidistributed buffer yields a higher throughput than with the central buffer. The situation is reversed however when F decreases. Indeed, below a critical F^*, the central buffer configuration is favourable. As the throughput depends monotonically on the parameter F, there exists a critical value F^* where the throughputs coincide (see Figure 5.5). In Figure 5.6, the location of the critical point F^* is calculated for varying indisposability factors.

Remember that the previous results are obtained in the hydrodynamic framework and hence for very small buffer capacities the expressions are to be interpreted with some care. In [DV], the errors made by taking the continuous limit in the dynamics have been analysed. In particular, it is shown how, by suitable buffer size renormalizations, upper and lower bounds of the throughput for the discrete dynamics can be obtained in the hydrodynamic framework.

Before closing this section, it necessary to emphasize that all the calculations we have performed did not take into account the time needed to transfer the parts to and from the buffers. Here again, a suitable renormalization of the buffer capacity enables one to fully take into account these effects [CM].

5.9 Summary and Conclusions

The dynamics of the population level of a buffer, when approximated by a continuous variable, can be described by stochastic differential equations in which the noise is continuous Markov chains. We have restricted our analysis to the Markov dynamics in which the time between failure and the time to repair are exponentially distributed. For non-Markovian situations, the same framework can be applied by introducing phase-type distributions for the random failing and repair times. In this framework, the Chapman-Kolmogorov equations which

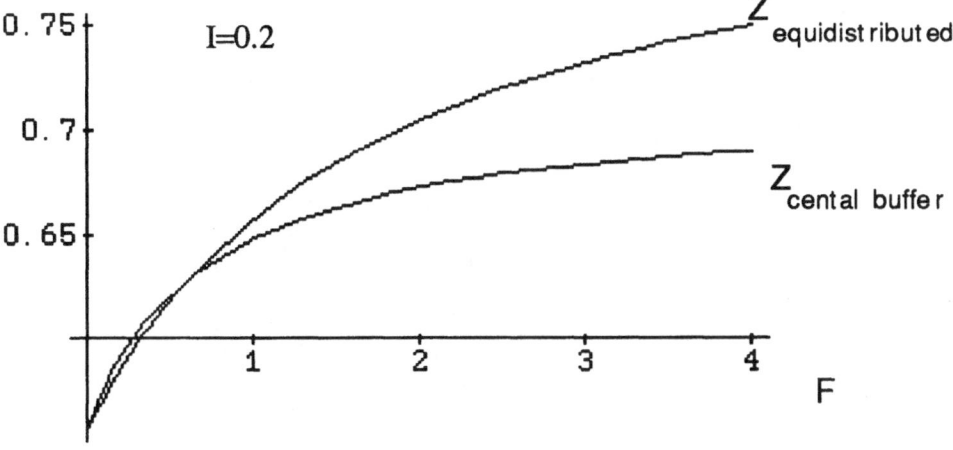

Fig. 5.5. Comparison of the throughput for equidistributed and "central lung" configurations.

describe the transition probability densities are high-order, hyperbolic, partial differential equations. Before the characteristics of the hyperbolic Chapman-Kolmogorov have reached the boundary of the buffer, the complete time dependent analysis is performed. For asymptotically large times, the boundary conditions (i.e. empty or filled buffer states) have to be treated with care as they directly influence the throughput of the line. Two failure mechanisms have been analysed, namely the OD (operation dependent failure proned machines), which characterizes machines which can fail only if operating (i.e. a blocked or a starved machine cannot fail) and the TD (time dependent failure-prone machines), for which failures can occur even if the machine is not producing but potentially able to produce, (i.e starved or blocked machines can fail). Finally, we briefly reviewed the aggregation method needed to deal with several working heads. An elementary example with four identical machines exhibits the complex behaviour of the throughput of the system. Indeed, we observe that not only the size of the buffers is important but also their locations in the line.

We have implicitly asssumed in this section that the production line is nearly balanced. Roughly speaking, the balancing of the line means that every isolated machine works with an identical mean throughput. If this is not the case, it is intuitively clear, that the formation of bottle necks will cause the system to be a slave of the performance of its poorest machine. In the context of balancing problems, let us mention recent studies of more flexible transfer mechanisms formalized in [BT]. In these system, inherent non-linearities of the dynamics

$I = \lambda/\mu$

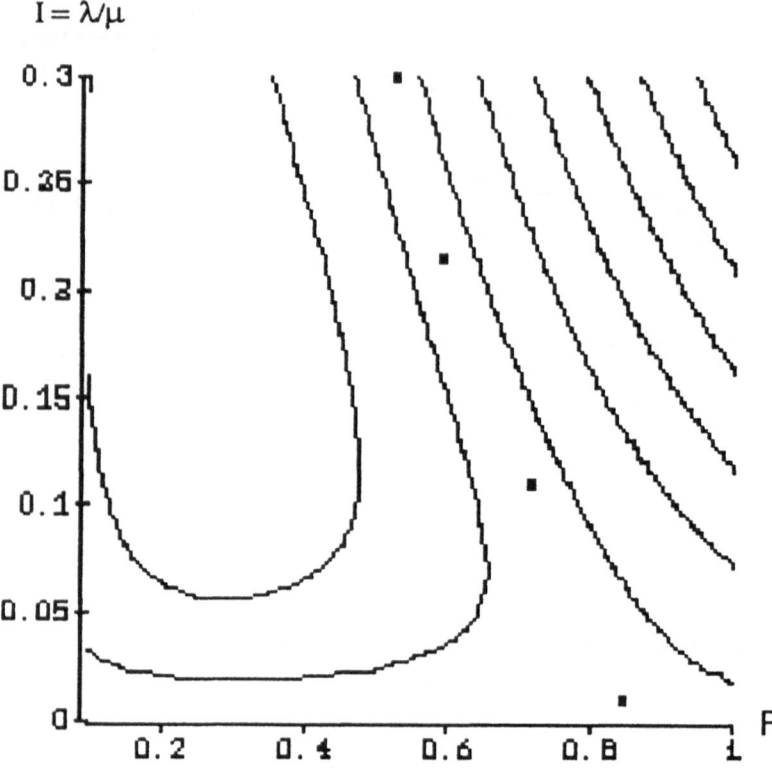

Fig. 5.6. Critical values F^* (small black squares) as a function of the in-disposability factor I. The drawn lines are equipotentials of the function $Z_{diff}(F, I) = Z_{TD}(F, I) - Z_{OD}(F, I)$.

governing the flux of parts enable the line to itself reach the balanced regime.

In this chapter, we have only considered a modeling in which the noise source is a continuous time Markov chain. A complementary approach to stochastic buffered flows using Brownian motion processes as noise sources is presented in [HR].

6 General Conclusions

The conception of an automatic production line (APL) or the improvement of an existing one, requires a vast capital of know-how that only experienced production engineers can provide. As globally this know-how can only be accumulated by spending years in specific environments, one could raise the question whether a part of this knowledge could be acquired and retransmitted by complementary means. A positive response to this question necessarily relies on the existence of a systematic approach to the problem of the conception of APLs. The need for this systematization has stimulated research activity directed towards the modeling of the behaviour of APLs. The present lecture notes attempt to contribute to this general effort and therefore do not aspire to be substitutes for solid engineering experience and experimental know-how.

The proposed guided tour through the world of APLs, illustrates the central role played by random phenomena in describing their dynamics. Whatever the factories of the future will be, stochastic processes will definitely influence both their conception and their operating policies. Several devices exist only in the realm of random phenomena of which part-orienting devices and buffers are two among a wealth of other possible examples.

Among the aims of the modeling of an APL, we mention the optimization of the fluxes of components and the estimation of its throughput and therefore the identification of relevant control parameters on which this quantity is dependent. In particular, the degree of sensitivity of the dynamics of APLs upon modifications of the external control parameters is particularly crucial. Indeed, the complexity of the present installations results in high building costs and forces engineers to improve the flexibility of the production apparatus. By flexibility it is understood that rather than being specialized to one item, the APL should be able to process a family or category of products. Changing the characteristics of the production induces modifications to the control parameters. To maintain the line operating with these changes, the range of parameters tuning the production fluxes, has to be far from the instability points of the underlying dynamical system. Furthermore, local fluctuations have to be quickly absorbed and should not leave the production process unduly perturbed. In other words, the production line dynamics has to present a character of generic stability. While the effect of fluctuations often has to be overcome for the completion of a specific task, situations also exist for which the presence of noise is beneficial. An illustration of this arises in the randomization of robot tasks. Mechanical randomization, used as an insertion tool, perfectly compensates for unknown model errors and imprecise sensing, while maintaining an overall simplicity. This procedure is essentially a mechanical offspring of the celebrated Monte-Carlo procedure of numerical analysis.

Random behaviours, non-linearities, generic stability are basic concepts inherent to the modern approach to dynamical systems. While so far largely unexplored, the analysis of APLs in the light of these concepts will without doubt help to achieve further progress toward a unifying general picture.

Acknowledgements.

Profs. S. Albeverio, Ph. Blanchard and L. Streit are warmly thanked for their invitation to the BiBoS Research Center of the University of Bielefeld. In particular, Ph. Blanchard encouraged me to complete these lecture notes. I have also benefited from numerous interesting discussions over the years with Profs. J. Figour and C. W. Burckhardt from the Ecole Polytechnique Fédérale de Lausanne, (E.P.F.L.), with Prof. M. Bétemps and Dr. F. Badano at the Institut National des Sciences Appliquées, (I.N.S.A.) of Lyon and Prof. J.-M. Proth and Dr. X.-L. Xie at the I.N.R.I.A. of Metz. I thank warmly Dr. B. S. Stevens for his numerous relevant suggestions, Mr. R. Romanowicz for the proofreading of the text and Mr. W. Maeder for his help in the drawings.

These notes could not appear without the support of my wife Cam Van and the patience of my children Clément and Isadora.

7 References

[AL]   Aldous, D.: Random walk on finite groups and rapidly mixing Markov chains. Séminaire de probabilités *XVII*, Lect. Notes in Math. **986**, 243-297, Springer.

[AR]   Arnold, V.I. and Avez A.: *Ergodic problems of classical mechanics.* (1968), Benjamin.

[BA1] Bapat, C. N., Sankar, S. and Popplewell, N.: Repeated impacts on a sinusoidally vibrating table reappraised. J. Sound and Vib. **108**, (1986), 99-115.

[BA2] Bapat, C. N. and Popplewell, N.: Several similar vibroimpact systems. J. of Sound and Vib. **113**, (1987), 17.

[BD1] Badano, F., Bétemps, M. B., Redarce, T. and Jutard, A.: Robotic assembly by slight random movements. Robotica **9**, (1991), 23-29.

[BD2] Badano, F., Bétemps, M. and Jutard A.: Assemblage automatisé par recherche aléatoire. APII (RAIRO) **26**, (1992), 35-40.

[BO]   Boothroyd, G., Poli C., and Murch L.E.: *Automatic Assembly.* (1982), Marcel Dekker, New-york

[BR]   Barraquand, J. and Latombe, J.-Cl.: Robot motion planning: A distributed representation approach. Int. J. of Robot. Res. **10**, (1991), 628-649.

[BT]   Bartholdi, J. J. and Eisenstein D. D.: A production line that balances itself. Preprint School of Industrial systems Engineering, Georgia Institute of Technology, Atlanta, (1993).

[BU]   Buzacott, J. A.:Automatic transfer line with buffer stocks.Int. J. Prod. Res. **5**, (1967), 182-200.

[CA]   Carvalho, J. C. M.:Transporteurs vibrants à déplacement imposé: Modélisation et applications. Thèse de l'Universsité de Franche Comté, Besanqn, (1991).

[CD]   Chandrasekhar, S.: Stochastic problems in physics and astronomy. Rev. Mod Phys. **15**, (1943), 1-89.

[CH]   Chow, W. M.: *Assembly Line Design. A methodology and applications.* (1990). Marcel Dekker. New-York.

[CI]   Ciesielski, Z. and Taylor, S. J.: First passage times and sojourn times for Brownian Motion in space and the exact Hausdorff measure of the sample path. Trans. Am. Math. Soc. **87**,(1963), 434-450.

[CL]   Coillard, P. and J. M. Proth, J. M.: Sur l'effet de l'adjonction des stocks tampons dans une fabrication en ligne. Rev. Belge Stat, Inf. et R.O. **24**, (1983), 1.

[CM] Commault, C. and Semery, A.: Taking into account delays in buffers for analytical performance of transfer lines. IIE Trans. **22**, (1990), 133.

[CW] Chu, W. W.: A mathematical model for diagnosing system failure. IEEE trans. Electronic Computers, **16**, (1967),327-331.

[CX]   Cox, D. R.: *Renewal theory.* (1962). Methuen.

[DA]   Dallery, Y. and Gerschwin, S. B.: Manufacturing flow line systems. A review of models and analytical results.Queuing Systems: Theory and Appl. (QUESTA), **12**, (1992), 3-94.

[DB]   Dubois, D. and Forestier J.P.: Productivité et en-cours moyens d'un ensemble de deux machines séparées par une zone de stockage. R.A.I.R.O. Autom. Syst. Analysis and Control **16**, (1981), 105-132.

[DI]   Diaconis, P.: *Group representations in Probability and Statistics.* I.M.S. Lect. Series **11**, (1988).

[DU]   Durret, R.: A new proof of Spitzer's result on the winding of 2-dimensional Brownian Motion. The Annals of Probab. **10**, (1982), 244- 246.

[DV]   David, R., Xie, X.-L. and Y. Dallery, Y.: Properties of continuous models of transfer lines with unreliable machines and finite buffers. IMA J. of Math. in Business and Industry, **6**, (1990), 281.

[ER]   Erdmann, M.: Randomization in robot tasks. Int. J. of Robot. Res. **11**, (1992), 399-436.

[FE]   Feigin, M. I.: Slippage in dynamics systems with collision interactions. Prikl. Mat. i Mekh., **31**, (1967), 559-562.

[FD]   Fedosenko, Yu. S. and Feigin, M. I. Periodic motions of a vibrating striker including a slippage region. Prikl. Mat. i Mekh., **35**, (1971), 845-850.

[FI]   Figour, J.: Assemblage automatisé. (1989), Cours EPFL (Lausanne).

[FI]   Figour, J.: Private communication.

[FO]   Forestier, J. P.: Modélisation stochastique et comportement asymptotique d'un système automatisé de production. RAIRO Automat. **14**, (1980), 127.

[GA]   Gardiner, G. W.: *Stochastic methods for physics, chemistry and the natural sciences.* (1983) Springer Verlag.

[GE]   Gershwin S. B. and Berman O.:Analysis of transfer lines consisting of two unreliable machines with random processing times. AIIE Transac. **13**, (1981), 2-11.

[GI]   Giraud, L., Ait-Kadi, D. and Guillot, M.: A hybrid model for fault diagnosis in manufacturing systems. Proceed. of the CAR's and FOF 8th Internatinal Conference on CAD CAM, Robotics and Factories of the Future (1992), 724-733. Metz.

[GO]   Goryunov, V.I., Dondoshanskaya, A. V., Metrikin, V.S. and Nagaev R.F.: Periodic motions of an object above a surface vibrating according to an anharmonic law.Prikl. Mekhanika **10**, (1974), 65.

[GR]   Gradshteyn,, I. S. and Stegun, I.M.: *Table of Integrals, series and products.* (1980), Academic Press.

[GU]   Guckenheimer, J. and Holmes, P. J.: *Non-linear oscillations, dynamical systems and bifurcations of vector fields.* Applied Math. Sciences **42**, (1983). Springer Verlag.

[HA1] Hänggi, P. and Talkner, P.: First passage time for non-Markovian processes. Phys. Rev. **A 32**, (1985), 1934.

[HA2] Hänggi, P.: Coloured noise in continuous dynamical systems: A functional calculus approach. in *Noise in Non-linear dynamical systems*, eds. I. F. Moss and P. V. E. Mc Clintock. Vol.1, 1989, 307-328.

[HE]   Heiman, M.S., Bajaj, A.K. and Sherman, P.J.: Periodic motions and bifurcations in dynamics of an inclined impact pair. J. of Sound and Vibrations **124**, (1988), 55-78.

[HF]   Hoffman, B., Pollack, H. and Weissman, B.: Vibratory insertion process: A new approach to non-standard component insertion. Robot. **8**, (1985), 1.

[HL]   Holmes, P.J.: The dynamics of repeated impacts with a sinusoidally vibrating table. J. Sound and Vibrtations **84**, (1982), 173.

[HN]   Han, I, Gilmore, B. J. and Ogot, M. M.: Synthesis and experimental validation of dynamic part-orienting devices. ASME, Design Eng. Advances in Design Automation, 44-1,(1992), 93.83-

[HO1]   Hongler, M.-O. and Figour, J.: Periodic versus chaotic dynamics in vibratory feeders. Hel. Phys. Acta. 62, (1989), 68.

[HO2]   Hongler, M.-O.: Stochastic dispersive transport: an excursion from statistical physics to automated production line design. Appl. Stoch. Models and Data Anal, 9, (1993), 139-152.

[HO3]   Hongler, M.-O., Badano, F., Betemps, M. and Jutard, A.: Random exploration approach for the automatic chamferless insertion4. Submitted to The Int. J. of Robotic Research (1994).

[HO4]   Hongler, M.-O. and Streit, L.: On the origin of chaos in gearbox models. Physica 29D, (1988), 402-408.

[HR]   Harrison, J. M.: *Brownian motion and stochastic flow systems.* J. Wiley, (1985).

[HS]   Horsthemke, W. and Lefever R.:*Noise-Induced Phase Transitions. Theory and Applications in Physics, Chemistry and Biology.* Springer (1984).

[IN]   Inoue,J., Miyaura, S., Nishiyama, A.: On the vibrotransportation and the vibro-separation. Bull. of the J.S.M.E. 11, (1968), 102.

[IT]   Itô, K. and Mckean, H. P.: *Diffusion processes and their sample paths.* Grundlagen der Mathematischen Wiss. in Einzeldarstellung 125, (1974), Springer.

[IV]   Ivanov, R.: Scanning Assembly. The Int. J. of Adv. Manufacturing Technology 4, (1989), 95.

[JA1]   Jaumard, B., Lu, S.-H. and Sriskandarajah C.: Design parameters for the selection and ordering of part-orienting devices. Int. J. of Prod. Research.28, (1990). 459.

[JA2]   Jaumard, B., Lou, S. X. C., Lu, S.-H. and Sriskandarajah, C.: Design of part-orienting Devices. Int. J. of Flexible Manuf. Syst. 5, (1993), 167-185.

[JE]   Jeong, K. and Cho, H.: Development of a pneumatic vibratory wrist for robotic assembly. Robotica 7, (1989), 9.

[JF]   Jafari, M.A. and Shanthikumar J. G. Exact and approximate solution to the two-stage transfer lines with general uptime and downtime distributions. IIE Trans. 19, 1987, 412-420.

[JI]   Jimbo,Y. Yokohama, Y. and Okabe, S.: Study of vibratory conveying. Int. Conf. on Production Eng. Tokyo (1974).

[JN]   Janicki, A. and Weron A.: *Simulation and chaotic behaviour of α-stable stochastic processes.* Marcel Dekker, (1993).

[KA]   Karlin, S. and Taylor, H. M.: *A first course and a second course in stochastic processes.* Acad Press (1975).

[KD]   Kendall, D. G.: Pole seeking Brownian Motion and bird navigation. J. of the Royal Stat. Soc. B36, (1974), 365-416.

[KE]   Ksendzov, A. A. and Nagaef, R. F.: Infinite-impact periodic modes in the problem of vibrating transport systems with tossing. Izv. Akad. Nauk SSSR, Mekhan. Tverd. Tela. 5, (1971), 29-35.

[KS]   Kemeny, J. G. and Snell, J. L.: *Finite Markov Chains.* Springer-Verlag (1976).

[KT]   Kent, J. T.: Eigenvalue expansions for diffusion hitting times. Z Wahrscheinlichkeitsth. verw. Geb. 52, (1980), 309.

[LA]   Lasota, A. and Mackey, M. *Probabilistic behaviours of deterministic systems.* (1985), Cambridge University Press.

[LI]   Lichtenberg, A. J. and Lieberman, M. A.: *Regular and stochastic motion.* Applied Math. Sciences 38, (1983). Springer Verlag.

[MA1] Maul, G. P. and Hildebrand, J. S.: Research for low cost flexible feeding of headed parts using bi-directional belts. Int. J.of Prod Res. 6, (1985), 1121.

[MA2] Maul, G. P. and Goodrich J. L.: A methodology for developing programmable feeders. I.E.E. Trans. 15, (1983), 330.

[MC] Minc, H.: *Non-negative matrices.* (1988), Wiley.

[ME] Messulam, P. and Yor, M.: On D. Williams pinching method and some applications. J. London Math. Soc. 26, (1982), 348-364.

[MI] Mitra, D.: Stochastic theory of a fluid model of producers and consumers coupled by a buffer. Adv. Appl. Prob. 20, (1988), 646.

[ML] Mello, T.M. and Tufillaro, N.B.: Strange attractors of a bouncing ball. Am J. of Phys. 55, (1987), 316-320.

[MO] Moon, F.C.: *Chaotic Vibrations.* (1987), Wiley.

[MV] Mevel, B. and Guyander, J. L.: Routes to chaos in ball bearings. J. of Sound and Vibrations 162, (1993), 471-87.

[MR] Marek, M. and Schreiber, I.: *Chaotic behaviours of deterministic dissipative systems.* (1991), Cambridge University Press.

[MS] Masoliver, J., Lindenberg, K. and West, B. J.: First passage for non-Markovian processes. Phys. Rev. A33, (1986), 2177.

[NA] Nagaev, R. F. : General problem of quasi-plastic impact. Izv. Akad. Nauk SSSR, Mekhan. Tverd. Tela. 3, (1971), 94-103.

[NE] Neuts, M. F.: Computational uses of the method of phases in the theory of queues. Comp. Math. Appl., 1, (1975), 151.

[OK1] Okabe, S., Kamiya, K., Tsujikado, K. and Yokoyoma, Y.: Vibratory feeding by non-sinusoidal vibratory optimum wave-form. Trans. of the ASME 107, (1985), 188.

[OK2] Okabe, S. and Yokoyoma, Y.: Study of a vibratory feeder with repulsive surface which has directional characteristics. Trans. of the ASME, j of Mechanical Design, paper presented at the Design Engineering Technical Conference, St. Louis (1979).

[PA] Paz, A: *Introduction to probabilistic automata.* Academic Press (1971).

[PE] Petersen, K.: *Ergodic theory.* (1983), Cambridge University Press.

[PI1] Pitman, J. and Yor, M.: Asymptotic laws of planar Brownian Motions. Ann. Prob. 14, (1986), 733-779.

[PI2] Pitman, J. and Yor, M.: Further asymptotic laws of planar Brownian Motions. Ann. Prob. 17, (1989), 965-1011.

[PK] Pinsky, M.: Differential equations with a small parameter and the central limit theorem. Z. Wahrschein. verw. Geb. 9, (1968), 101.

[RE] Redford A.H., and Boothroyd G.: Vibratory feeding. Proc. Instn. Mech. Eng. 182, (1967-68), 135.

[RI] Risken, H.: *The Fokker-Planck Equation.* (1989), Springer.

[RO] Rosenthal, J.S.: Markov chains, eigenvalues, and coupling. Technical reports No. 9320, (1993), Dpt. of Statistics, University of Toronto.

[RS] Rubino, G. and Sericola, B.: On weak lumpability in Markov Chains. J. Appl. Prob. 26, (1989), 446.

[RU] Ruelle, D.: *Chance and Chaos.* Princeton (1993).

[SA] Sancho, J. M., San Miguel, M., Katz, S. L. and Gunton J.D.: Analytical and numerical studies of multiplicative noise. Phys. Rev. A 26, (1982), 1589-1609..

[SC] Schweigert, U.: *Toleranzausgleichssysteme für Industrieroboter am Beispiel des feinwerktechnischen Bolzen-Loch-Problems.* (1992), Springer-Verlag.

[SE]   Sethi, S. P., Sriskandarajah, C. and Rao, M. R.: Heuristic for selection and ordering of part-orienting devices. Operation research **38**, (1990), 84-98.

[SP]   Spitzer, F.: Some theorems concerning the 2-dimensional Brownian Motion. Trans. Am. Math. Soc. **87**, (1958), 187-197.

[ST]   Seneta, E.: *Non-negative matrices and Markov chains.* (1981), Springer series in statistics, second edition.

[SV]   Sveshnikov, A.A.: *Problems in probability theory, mathematical statistics and the theory of random functions.* (1968), Dover.

[TA]   Taniguchi, O., Sakata, J., Suzuki, Y. and Osanai, Y.: Studies on vibratory feeder. Bull J.S.M.E. **6**, (1963), 37.

[TE]   Terracol, C. and David, R.: Performance d'une ligne composie de machines et de stocks intermédiaires. APII **21**,(1987), 239.

[TU1]   Tufillaro, N.B., Mello, T. M., Choi and A. M. Albano, V. M.: Period doubling boundaries of a bouncing ball. J. Physique **47**, (1986), 1477

[TU2]   Tufillaro, N. B. and Albano, A. M.: Chaotic dynamics of a bouncing ball. Am. J. of Phys. **54**, (1986), 939.

[TU3]   Tufillaro, N. B. and Abott, M.: Bouncing ball dynamics for the Macintossh Computer. (1988).

[VA]   Vaynkof, Va. F. and Inosov, S.V.: Non-periodic motion in vibratory conveying in states of operation accompanied by throwing. Mechanical Sciences.- Maschinovedeniye **5**, (1976),1.

[VE]   Veitz, V. L. and Beilin, I. Sh.: Dynamics of transportation of a material particle with stochastic characteristics along a horizontal plane. Mechanism and Machine Theory **7**, (1972), 155-65.

[VK]   van Kampen, N., G.: *Stochastic Processes in Physics and Chemistry.* (1981), North Holland .

[WA]   Warnecke, H., Frankenhauser, B., Gweon, D. and Cho, H.: Fitting of crimp contacts to connectors using industrial robots supported by vibrating tools. Robotica **6**, (1988), 123.

[WI]   Wiesenfeld, K. and Tufillaro, N. B.: Suppression of period doubling in the dynamics of a bouncing ball. Physica **26D**, (1987), 321.

[WG]   Winjgaard, J.: The effect of interstage buffer storage on the output of two unreliable production units in series, with different production rates. AIIE Trans.**11**, (1979), 42-47.

[WO]   Wood, L.A. and Byrne, K. P.: Analysis of a random repeated impact process. J. of Sound and Vibrations.**78**, (1981), 329-45.

[YE]   Yeong, M.,Ruff, L. and De Vries, W. R.: A survey of part presentation, feeding and fixturing in automated assembly systems. A.S.M.E. Design Engineering, **33**, (1991), 83-90.

[ZI]   Zimmern, B.: Etude de la propagation des arrêts aléatoires dans une chaîne de production. Rev. de Stat Appl. **4**, (1956), 85.

Springer-Verlag
and the Environment

We at Springer-Verlag firmly believe that an international science publisher has a special obligation to the environment, and our corporate policies consistently reflect this conviction.

We also expect our business partners – paper mills, printers, packaging manufacturers, etc. – to commit themselves to using environmentally friendly materials and production processes.

The paper in this book is made from low- or no-chlorine pulp and is acid free, in conformance with international standards for paper permanency.

Lecture Notes in Physics

For information about Vols. 1–394
please contact your bookseller or Springer-Verlag

New Series m: Monographs